Samples and
Standards

Wiley India Pvt. Ltd.

Analytical Chemistry by Open Learning

Project Director
BRIAN R CURRELL
Thames Polytechnic

Project Manager
JOHN W JAMES
Consultant

Project Advisors
ANTHONY D ASHMORE
Royal Society of Chemistry

DAVE W PARK
Consultant

Administrative Editor
NORMA CHADWICK
Thames Polytechnic

Titles in Series:

1. Samples and Standards
2. Sample Pretreatment and Separation
3. Classical Methods Vol. 1
4. Classical Methods Vol. 2
5. Measurement, Statistics and Computation
6. Using Literature
7. Instrumentation
8. Chromatographic Separations
9. Gas Chromatography, 2nd Edition
10. High Performance Liquid Chromatography, 2nd Edition
11. Electrophoresis
12. Thin Layer Chromatography
13. Ultraviolet and Visible Spectroscopy
14. Fluorescence and Phosphorescence Spectroscopy
15. Biological Applications of Infra Red Spectroscopy
16. Atomic Absorption and Plasma Spectroscopy, 2nd Edition
17. Nuclear Magnetic Resonance Spectroscopy
18. X-Ray Methods
19. Mass Spectrometry, 2nd Edition
20. Scanning Electron Microscopy and X-Ray Microanalysis
21. Principles of Electroanalytical Methods
22. Potentiometry and Ion Selective Electrodes
23. Polarography and Other Voltammetric Methods
24. Radiochemical Methods
25. Clinical Specimens
26. Diagnostic Enzymology
27. Quantitative Bioassay
28. Assessment and Control of Biochemical Methods
29. Thermal Methods
30. Microprocessor Applications
31. Environmental Analysis
32. Quality in the Analytical Chemistry Laboratory

Samples and Standards

Analytical Chemistry by Open Learning

Authors:
BRIAN W. WOODGET
Hatfield Polytechnic, UK

DEREK COOPER
North Staffordshire Polytechnic, UK

Editor:
NORMAN B. CHAPMAN

on behalf of ACOL

Published on behalf of ACOL, London
by
JOHN WILEY & SONS
Chichester · New York · Brisbane · Toronto · Singapore

Samples and Standards

Reprinted June 1991
Reprinted October 1995

Published by permission of the Controller of
Her Majesty's Stationery Office

Authorized reprint by Wiley India Pvt. Ltd., 4435/7, Ansari Road, Daryaganj, New Delhi – 110002.

Reprint: 2008

Printed at: *Ar Emm International, New Delhi*

ISBN: 978-81-265-1711-4

The ACOL Series

This series of easy to read, user-friendly texts has been written by some of the foremost lecturers in Analytical Chemistry in the United Kingdom. The texts are designed for training, continuing education and updating of all technical staff concerned with Analytical Chemistry.

These texts are for those interested in Analytical Chemistry and instrumental techniques who wish to study in a more flexible way than traditional institute attendance or to augment such attendance.

Analytical Chemistry by Open Learning, ACOL, provides training resources—books, courses and computer programs: Courses based on ACOL, incorporating practical workshops and tutor assessment, are provided by Thames Polytechnic in conjunction with the Royal Society of Chemistry. BTEC qualifications are awarded for successful completion of courses and lead to a diploma that would be an appropriate preparation for applicants to the Royal Society of Chemistry for Licentiateship status, LRSC.

Thames Polytechnic will continue to support these 'Open Learning Texts', to continually refresh and update the material and to extend their coverage. For further information on ACOL, the RSC courses and the BTEC diploma, please contact:

The ACOL Office,
Thames Polytechnic,
Avery Hill Road,
Eltham,
London,
SE9 2HB,
U.K.

How to Use an Open Learning Text

Open learning texts are designed as a convenient and flexible way of studying for people who, for a variety of reasons cannot use conventional education courses. You will learn from this text the principles of one subject in Analytical Chemistry, but only by putting this knowledge into practice, under professional supervision, will you gain a full understanding of the analytical techniques described.

To achieve the full benefit from an open learning text you need to plan your place and time of study.

- Find the most suitable place to study where you can work without disturbance.
- If you have a tutor supervising your study discuss with him, or her, the date by which you should have completed this text.
- Some people study perfectly well in irregular bursts, however most students find that setting aside a certain number of hours each day is the most satisfactory method. It is for you to decide which pattern of study suits you best.
- If you decide to study for several hours at once, take short breaks of five or ten minutes every half hour or so. You will find that this method maintains a higher overall level of concentration.

Before you begin a detailed reading of the text, familiarise yourself with the general layout of the material. Have a look at the course contents list at the front of the book and flip through the pages to get a general impression of the way the subject is dealt with. You will find that there is space on the pages to make comments alongside the text as you study—your own notes for highlighting points that you feel are particularly important. Indicate in the margin the points you would like to discuss further with a tutor or fellow student. When you come to revise, these personal study notes will be very useful.

Π When you find a paragraph in the text marked with a symbol such as is shown here, this is where you get involved. At this point you are directed to do things: draw graphs, answer questions, perform calculations, etc. Do make an attempt at these activities. If necessary cover the succeeding response with a piece of paper until you are ready to read on. This is an opportunity for you to learn by participating in the subject and although the text continues by discussing your response, there is no better way to learn than by working things out for yourself.

We have introduced self assessment questions (SAQ) at appropriate places in the text. These SAQs provide for you a way of finding out if you understand what you have just been studying. There is space on the page for your answer and for any comments you want to add after reading the author's response. You will find the author's response to each SAQ at the end of the text. Compare what you have written with the response provided and read the discussion and advice.

At intervals in the text you will find a Summary and List of Objectives. The Summary will emphasise the important points covered by the material you have just read and the Objectives will give you a checklist of tasks you should then be able to achieve.

You can revise the Unit, perhaps for a formal examination, by re-reading the Summary and the Objectives, and by working through some of the SAQs. This should quickly alert you to areas of the text that need further study.

At the end of the book you will find for reference lists of commonly used scientific symbols and values, units of measurement and also a periodic table.

Contents

Study Guide

This Unit is rightly the first in the series 'Analytical Chemistry by Open Learning', in that it's intention is to set the scene on the subject, so that the more detailed scientific parts which come later in the series will acquire a logical and intellectual coherence. The unit can be divided into three separate topic areas, each of which may be studied separately if you wish. The topic areas are;

Part 1	The Analyst's Approach
Parts 2, 3, 4	Sampling
Parts 5 to 11	Standardisation and Calibration

All of these topics are fundamental to a good understanding of analytical chemistry, and a knowledge of them is essential for all who aspire to be called 'Analytical Chemists'.

As this may well be your first venture into open learning, possibly you will find it difficult initially to come to grips with the style. You must realise however that students of 'Analytical Chemistry by Open Learning' come from a wide variety of backgrounds of experiences, and the text is presented in such a way as to be understandable by all.

Part 1 of this Unit shows how analytical chemistry and some aspects of the wider field of analytical science relate to other areas of scientific and commercial activity. Questions like 'why is the analysis to be carried out?', 'is the analysis really necessary?', 'what do the results mean?' and many others, are all questions that the analytical chemist has to ask either before embarking on an analysis, during an analysis, or after the results from an analysis have been obtained. This Part of the Unit sets out to explain the questions that the analytical chemist should pose, and attempts to show some of the ways they can be answered as well as some of the pitfalls which may occur in practice. The analyst's approach is illustrated by several case studies which set the context for analysis and form an important part of this introduction.

The importance of taking, and analysing a representative sample cannot be stressed too strongly. Parts 2, 3 and 4 of this Unit consider the principles and theory of sampling, and go on to describe methods by which the samples may be taken and to describe typical sampling apparatus used for the sampling of solids, liquids, and gases in static and flowing situations. In any discussion of sampling it is necesary to use statistical methods to assess the precision and accuracy of a sampling procedure. Part 3 attempts to explain briefly some of the statistical theory required for the Unit but the learner is advised to refer to the Unit of this ACOL series: Measurement, Statistics and Computation, for an in-depth study of statistics.

The remainder of the Unit deals with methods of calibration, preparation of standards and evaluation of analytical procedure. As the majority of instrumental methods are comparative (i.e. require calibration with standard substances), the results obtained by these methods can be no better than the quality of the standards used and the care taken in producing the calibration. Consideration is given therefore to the choise of standard substances, the use of reference materials and methods of checking analytical performance.

Some parts of the Unit are, of necessity, quantitative, and include a number of mathematical derivations. Many of you will be familiar already with the use of calibration graphs, and the application of standard addition and internal standard procedures in instrumental analysis. But how many of you are aware of the mathematical background underlining these quantitative procedures, and the assumptions which are made when they are applied? The derivations and other mathematic manipulations are on the whole simple, requiring a level of mathematical ability no greater than O'level or equivalent. The text is supplemented with 'real' examples, to illustrate the application of these procedures, together with self-assessment questions to test your knowledge and understanding. You are advised to approach these quantitative questions, by imagining yourself performing the analysis described, rather than considering them as a piece of simple arithmetic. In this way, you are more likely to remember for instance, dilution factors, which must be taken into account if the correct result is to be obtained.

Bibliography

If you would like to pursue some of the topics introduced in this Unit, the following books would be useful sources of information:

For Part 1, the analyst approach:

– G. E. Baiulescu, C. Patroescu and R. A. Chalmers, *Education and Teaching in Analytical Chemistry*, Ellis Horward, 1982.

For Parts 2 to 4, sampling:

– C. L. Wilson and D. W. Wilson, etc, *Comprehensive Analytical Chemistry*, Vol 1a, chapter 3, Elsevier, 1959

– D. C. Cornish, G. Jepson and M. J. Smithwaite, *Sampling Systems for Process Analysis*, Butterworths, 1981.

For Parts 5 to 11, standardisation and calibration:

– *Official Standardised and Recommended Methods of Analysis*, The Royal Society of Chemistry, 1973.

– *British Pharmacopoeia*, The British Pharmacopoeia Commission

– *Official Methods of Analysis of the Association of Official Analytical Chemists*, 14th edn., 1984.

1. The Analyst's Approach

1.1. THE PURPOSE OF ANALYSIS

Let us start by asking the question:

'What is analytical chemistry?', or if you prefer it

'What is analytical chemistry about?'

Π As a student of a course in analytical chemistry you might assume that the question is trivial, or possibly rhetorical. Before reading on, just have a go at a reply!

If you try to formulate a response it is likely that you will end up in one of two general areas.

One type of response deals with the scientific and technical aspects of the measurement of compositional and constitutional features of samples. In some ways this is a layman's view of the chemical analyst; someone who identifies unknowns, and produces the percentage composition of known materials. The activity is perceived in the *context of the results*.

The second type of response sees analytical chemistry as the provider of one class of data-input into some decision-making process. That is, the significant activity and the results are always viewed within the *context of their further use*.

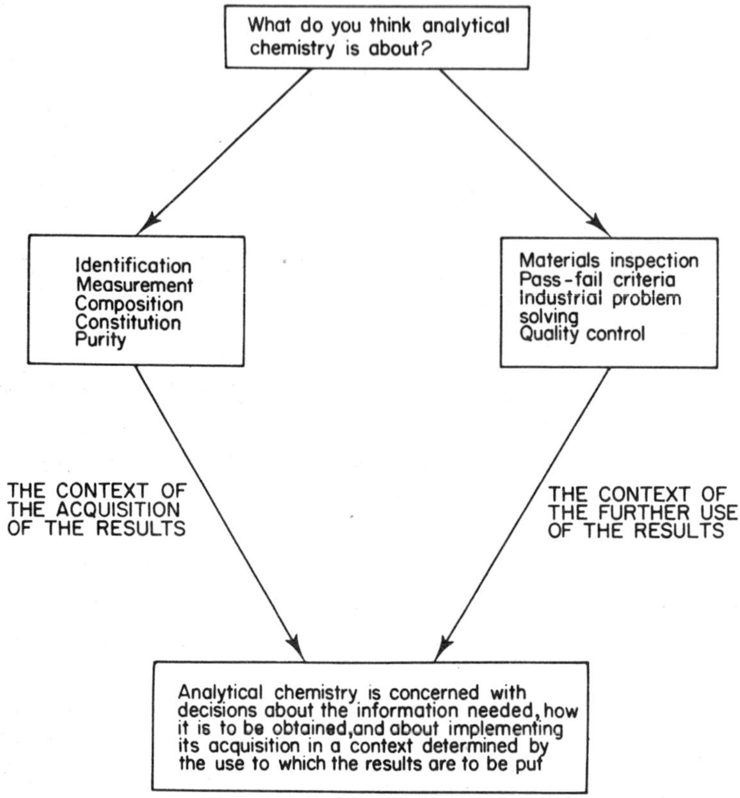

Fig. 1.1a. ***The context of chemical analysis***
Unfortunately many of us have been programmed to think of the left-hand side only

If your response to our opening question included both aspects you are already well on the way to attaining the objectives which are central to this introduction. We need to maintain a broad utilitarian view of chemical analysis, even when we become deeply embedded in any of the highly specialised fields covered by later sections. Thus the analyst is concerned not simply with the acquisition of analytical results but also with the decisions for which his results will be an input. Do not forget that however much you may weigh things in the balance, perhaps in a literal sense, your results will also be weighed in the balance of their *usefulness* to someone else. Cherished concepts of, for example, *accuracy* and *precision* become important only if they are useful, that is if they are relevant to a decision to do or not to do something, to accept or reject something, to change public policy, or to continue as before.

Analytical chemistry has been traditionally associated with the twin skills required for providing information on the *identification* and *composition* of materials, loosely called samples. This has led to the division of the subject area into *qualitative* analysis (ie what is it?) and *quantitative* analysis (ie how much of each component is there?). Two decades ago many analytical methods were limited to quantitative analysis at the level of 'percentage composition' of samples and it was feasible to attempt a 'total analysis'. However, during the last twenty years there has been such a rapid growth of technical expertise in both analytical areas and in associated fields that the lower limits of detection and measurement have been moving constantly downwards. After being satisfied with 0.1% as a lower limit for many analytes, we are now able routinely to obtain results at the parts per million (ppm) or even the parts per billion (ppb, ie 1 in 10^9) level.

There are, of course, many analyses for which precision at the 1% level is still quite satisfactory. However, this extended range of possibilities has forced the analyst to look more carefully 'upstream'. That is, we have to become increasingly aware of the treatment our samples receive before they undergo the actual analytical measurements. For example, we might need to know exactly how a sample was collected or what treatment it underwent during concentration procedures. Indeed the analyst may insist that the whole process of collection, separation, concentration, measurement, and interpretation are under his control, and done in a standard and agreed way. You will find that analysts refer to this standard and predefined way as a *protocol*.

Also, the massive array of analytical techniques now at one's disposal, and the speed with which vast quantities of results can be generated are obliging analysts to look 'downstream'. That is, they need to know exactly the use to which their results are to be put. This is to ensure that appropriate protocols can be designed for the use envisaged. A further development which is linked to the increasing skills of the analyst is an increasing move towards the determination of one or more components in *many* samples rather than the *total* analysis of a smaller number of samples.

We should see chemical analysis as a *whole process* stretching from collection, through pre-treatment, acquisition of results (the traditional view of analysis) to data processing and evaluation. Fig. 1.1b illustrates this.

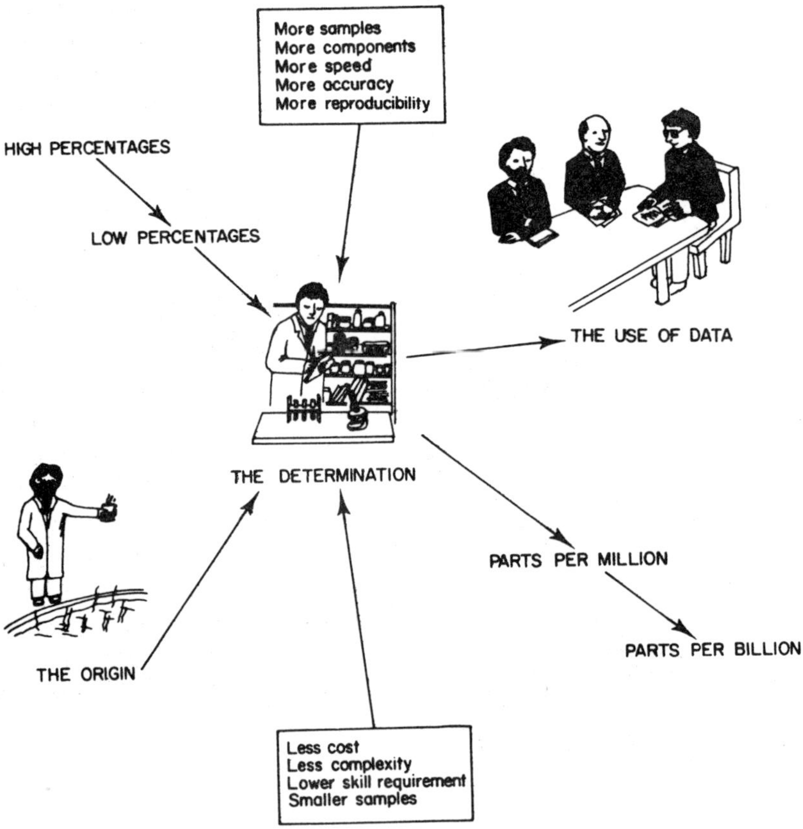

Fig. 1.1b. ***The scope of the analytical protocol***
Chemical analysis should be seen as a 'whole process', controlled by a defined protocol related to objectives derived from the eventual application

But who selects the analyte, ie the 'thing being analysed for'?

This question is, of course, the vital link between the user with his application problem and the analyst involved with the steps we have just listed. Like all interesting questions there is no clear once-and-for-all answer, but rather a fuzzy range of possibilities; we shall be exploring these in the next few pages.

We can illustrate the need for this whole-process view by relating a fictional and extreme case from an old cloak and dagger movie. The plot centres on a letter, it becomes important to know something about this letter and it is duly '... sent to the lab for investigation'. After some careful and time-consuming work the chemist is able to report that it has been exposed to a fairly concentrated sodium chloride solution.

'We could have told you that ourselves', reply our heroes, 'after all we fished it out of the sea!'

'Pity you didn't tell me earlier' mutters the disgruntled analyst.

'Well has it been opened or not?'

'Of course it has, five minutes under the microscope would have told you that!' replies the analyst triumphantly, honour duly satisfied.

Although this is entirely fictional the point which it makes is not as far-fetched as all that, as you will see from some of the cases we shall examine later in the introduction.

A diagrammatic representation of this broadened concept of chemical analysis helps to make the point. In Fig. 1.1c, although we readily see the strong clockwise move from 'noon' to 'midnight', we should also note the informed feed-back between the compartments, represented by the double arrows.

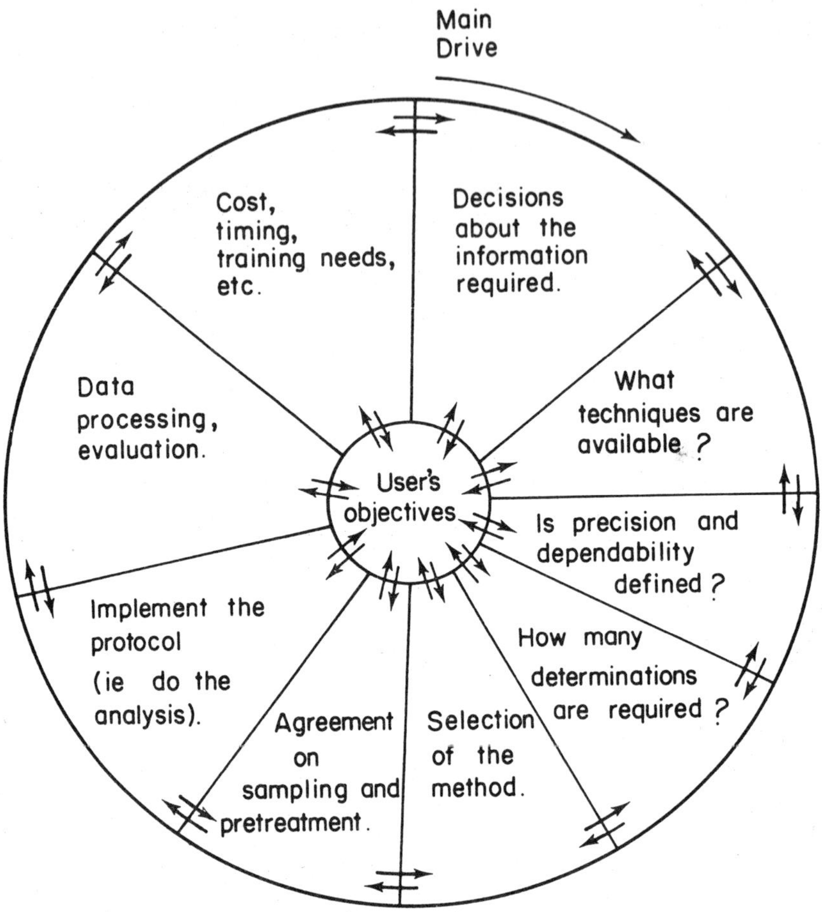

Fig. 1.1c. ***The broader concept of analytical chemistry***
Traditional discipline boundaries are not helpful to the analyst

1.2. STEPS IN A CHEMICAL ANALYSIS

In this Section we expand the general features shown in Fig. 1.1a as part of our strategy for making the *context* of the analysis more intelligible.

Let us start with a question: Can the analyst 'state the problem'?

The general view is that if the problem can be stated completely it probably can be solved. It is usually better to try to state the problem in non-analytical terms before translating it into analytical language. Then with a knowledge of the nature of the sample, the chemistry and physics of the methods, and the limits and practicability of the instrumentation, we would hope to proceed. This may be by using a well-validated and established method, or if necessary by developing a method and evaluating it; in both cases this is followed by evaluation of the results.

As you proceed with subsequent Units of this series, remind yourself constantly of this view of the chemical analyst and be at pains to distinguish your view of your own goals from the commonly held idea of a knob-twisting determinator or 'cookbook' analyst.

A short case-study illustrates some of the points made so far. As you read the case-study remember that we are interested in the flow of the ideas rather than the details of the chemistry.

1.3. CASE-STUDY: A PROBLEM OF SLUDGE

A moderate-sized chemical firm manufactures toluene-*p*-sulphonic acid by the one-step reaction of toluene and oleum. After the reaction, the mixture is diluted with an equal volume of water to cool it. The product is then transferred to road tankers, in which it is shipped to various customers in the metal-finishing business. Without warning, tanker loads are rejected by customers on an apparently random basis, because of the appearance of an unidentified sludge in the product.

The production manager brings a sample of the sludge to the analyst along with a sample of the mains water. The latter is blamed by the shift charge-hand for the problem, as the Water Authority have recently changed the supply to a hard-water source.

What do you as the analyst do? Let's have a look at some of the options.

(*a*) Quantitative analysis for Ca^{2+} and Mg^{2+} in the water.

(*b*) Quantitative analysis for Ca^{2+} and Mg^{2+} and perhaps other ions in the sludge.

(*c*) Qualitative analysis of the sludge.

(*d*) Send the plant manager back and say that it is standard practice for the analyst to collect all samples.

(*e*) Have a discussion with the plant manager about the operation of the process.

(*f*) Order the analysis of all starting materials for any impurities.

(*g*) Advise a preliminary look at the delivery schedule for starting materials and look for coincidence between 'bad' samples and deliveries.

(*h*) Other options of your own devising!

You can see that the choice of possible action begins to expand. Not only are some of the options barely analytical but some like 'analyse the starting materials' are capable of a very wide interpretation. Does the plant manager really want an identification of impurities whatever the level? Does he mean to the nearest 1%, 0.1% or ppm, and what is more, is he likely to know what he really needs?

We must ask ourselves a more fundamental question. What is the plant manager's real problem, for which he is seeking an analytical solution?

His problem is essentially commercial, ie his customers are rejecting batches of his product. This in turn is leading to added transport costs, extra production costs to meet the short-fall, storage costs for the rejected material, and bad publicity. His problem is decidedly *not* analytical, so the analyst must be wary of allowing him to set the analytical objectives.

You can see that the analyst must interpret his role within the context of the usefulness of his results to someone else, in this case the plant manager. The way in which samples are taken, the decisions about quantitative and qualitative analysis, and the acceptable levels of accuracy and reproducibility are never absolutes. They are not drawn out of some isolated view that 'this is the standard method for such and such'. These things must always be related to the eventual use of the results.

What analytical information is actually needed? This is always the most important question which precedes the actual collection and analysis of samples. The answer, of course, varies from application to application; in this example we need to know what the impurity is, then we can ask how it got there and how its presence can be avoided.

In this case qualitative analysis identified the sludge as iron(II) toluene-*p*-sulphonate, ie the iron(II) salt of the product itself. Has the analyst finished his job? After all, he can tell the plant manager what the results of the analysis are. We realise that that is not the plant manager's problem, so what else can the analyst do?

We could carry out a qualitative analysis of all starting materials with the specific objective of detecting iron as an impurity. This is in fact quick and cheap to do. When this was done, it turned out that iron was present in all the starting materials, and indeed in the cooling water, as well as being present in both acceptable and rejected batches of the product. In fact iron(III) is a common impurity in many bulk industrial chemicals.

The analyst readily recognised the difference between iron(II) and iron(III) and suspected some form of disturbance in the equilibrium. Separate qualitative analysis specifically for iron(II) revealed that this was present only in the rejected product. Something in the reaction must have been producing reducing conditions. A little chemistry identified the cause. When the reaction was carried out by adding oleum to toluene there was no local excess of oleum and product batches were good. When the addition was in the reverse order, local excesses of oleum (saturated with SO_2) produced the reducing conditions which gave rise to the iron(II) salt which happened to be less soluble than the iron(III) analogue, and so came out as a sludge.

As well as being able to tell the plant manager what his impurity was, the analyst was able to tell him why it was there and advise on the way to avoid the problem.

What have we learned from this example?

(*a*) It is not always clear at the outset what analytical information is actually necessary.

(*b*) The precision required is not always apparent at the outset and is not always high.

(*c*) The user's interest in precision extends only to solving his problem.

(*d*) Users of the results may not be the best people to decide what is necessary.

(*e*) At times unsophisticated, almost crude causes can give rise to commercial disasters.

1.4. THE PLAN FOR A CHEMICAL ANALYSIS

In the next few paragraphs we shall approach the idea of a plan for an analysis by asking a series of questions. At the first reading you should regard these as somewhat rhetorical: they are intended to provoke thought rather than to stimulate a specific answer.

Can you use the case we have just examined to develop a plan? At the moment we have the one question. Try a few of your own ideas as the next phases of the plan, Fig. 1.4a.

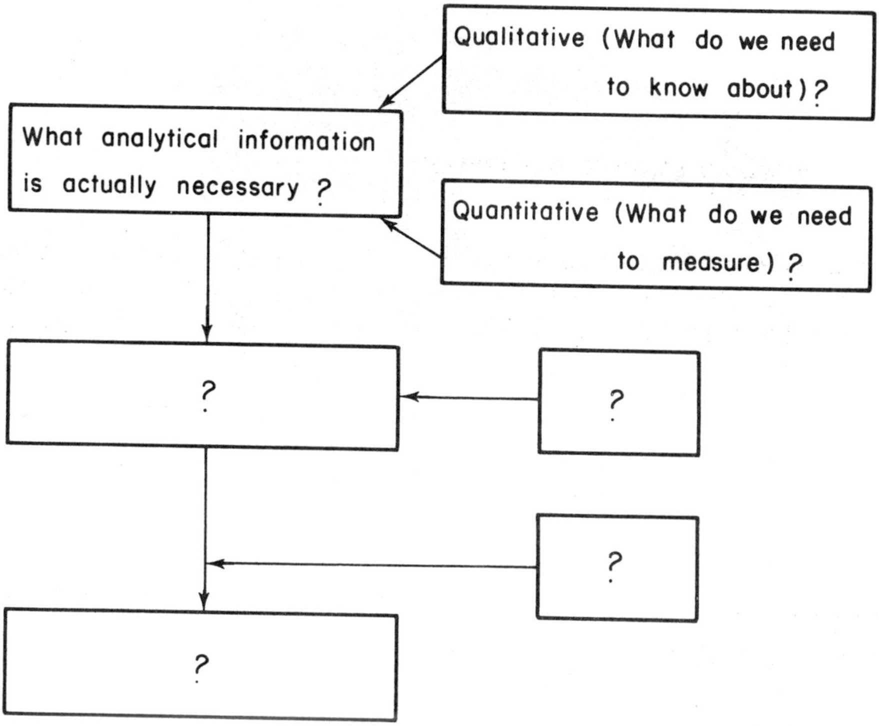

Fig. 1.4a. ***The analytical plan, a beginning***
What other questions should we be asking?

Once we have decided what to look for and what to measure we have to ask the next question.

What techniques are available to do this?

There are, of course, many texts and manuals on the subject of analytical chemistry, as well as a rich field of research literature. If you were, for example, considering the question '... how do I determine boron?' you would be quite likely to turn up at least a dozen methods in the literature. This is clearly not of much use to us, but luckily some of these techniques will be ruled out by our next question.

Has our application defined the required levels of general dependability, ie accuracy and reproducibility, and if so, how do these compare with the levels available?

This question will eliminate a number of one-time contenders. If, for example, you need to determine PCBs (polychlorinated biphenyls) in soil samples, it is no good looking towards a method which is sensitive only above 0.01%, as your range of interest will be in the ppb range. Equally it is pointless carrying out a determination of, say Vitamin C, in items of commerce to a precision of 0.0001%, when most of the controls are in hundreds of milligrams per hundred cm^3 of sample.

How many determinations are to be carried out?

The fundamental point here is to ask how much additional information is added by each extra determination, compared to the extra cost incurred in producing the extra results. There are, of course, two approaches to the number of analyses executed: one is how many samples should we collect and the other is how many measurements should we make on any one sample? In some cases each determination may cost the same amount, whereas in others the cost may be largely in setting up a method, and the additional cost of further determinations may be trivial. The answer, in any particular case, will of course depend on the application. Indeed it may sometimes be necessary to resort to statistical methods to answer this question.

What method should be used?

Our series of questions is now beginning to focus upon something which looks like the traditional idea of chemical analysis. There are, of course, many different methods available for the immense range of potential analytes. The actual selection of these will be dealt with in the later Units of this series. Our intention here is to comb out the common features which should aid our selection. These will be things like the time-scale required for a result, the number of samples and their temporal distribution (eg 50 at one sitting or one a week for a year), the cost per analysis, cost per additional analysis, availability of equipment and personnel, potential variability in the sample, possible interferences in the method, and so on.

Π Have another look at our diagrammatic approach to the plan, Fig. 1.4a, and add a few more boxes if you like.

Your diagrammatic plan should now look something like Fig. 1.4b.

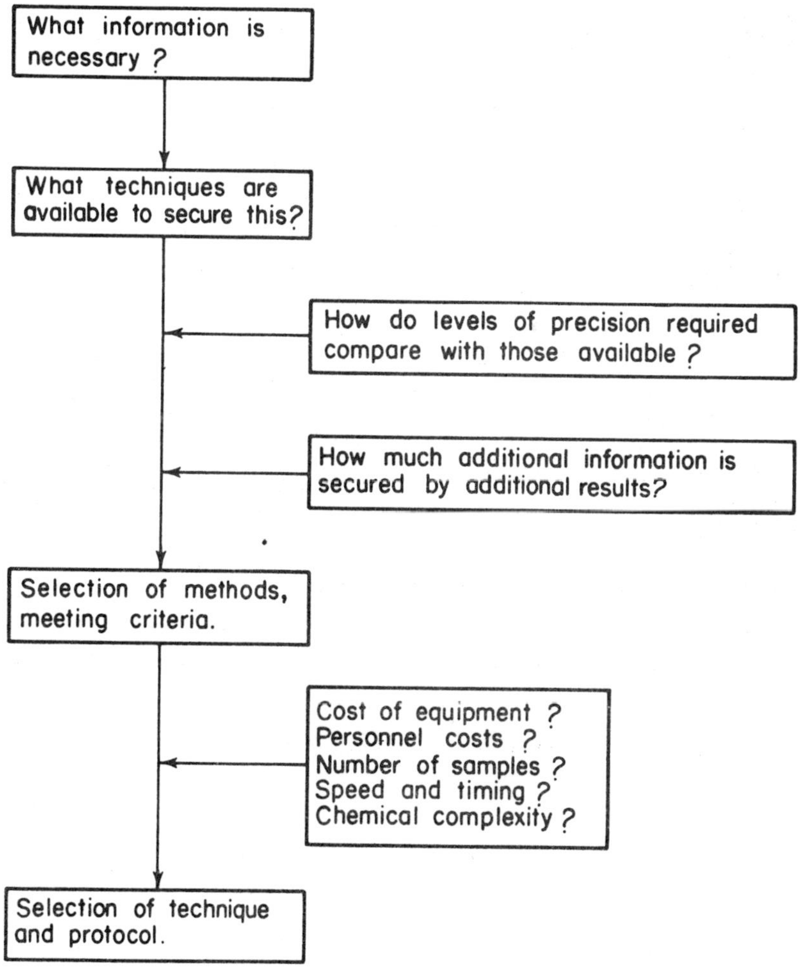

Fig. 1.4b. ***The analytical plan***
How the appropriate techniques are identified

1.4.1. Taking Samples

At this stage we have identified the technique to be used; we now have to consider obtaining the sample and preparing it for analysis. Our main problem is that of obtaining a laboratory-sized sample which is *representative* of the bulk material being investigated.

Thus the analyst needs to know something about the bulk analyte: is it homogeneous or heterogeneous, or even quasi-homogeneous (apparently homogeneous but not in fact so on a micro-scale)? We usually assume that gas mixtures are homogeneous, eg in air-pollution measurements, but a sample of soil is quite another thing. It is quite clearly heterogeneous. Many modern techniques require only the minutest of samples to be actually presented to the analytical instrument; you can readily see the problem with a heterogeneous sample. One has to be careful that non-representative weighting, because of particle size differences, does not occur either by segregation in processing or by preferential uptake during the measurement. Even quasi-homogeneous things, for example milk, can present problems for the unwary: we all know that milk separates on being kept! Therefore you can see that the analyst has to be aware of the possibility of unsuspected gradients in the field sample, and of the possibility of fractionation of the sample induced by the actual collection and subsequent treatment it receives.

At their introduction to analytical chemistry many chemists may be tempted to feel that the analytical establishment overdoes the importance of sample collection. 'After all', they may say, 'the real science takes place in the laboratory, doesn't it?' However, the following should be indelibly etched in your mind, or at least framed and hung up in your laboratory!

PRECISE ANALYTICAL MEASUREMENTS ON A NON-REPRESENTATIVE SAMPLE ARE A WASTE OF TIME.

For example, if the sampling error is ± 10 ppt (parts per thousand), even laboratory procedures which are precise to ± 10 ppb (parts per billion) do *not* make the result better than ± 10 ppt. The variety of questions we should now ask is expanding. The detailed answers to these problems of sampling are dealt with in later parts of this Unit but the list below illustrates the scope for getting off on the wrong foot.

- Should combined samples be used?
- Should quartering methods be used?
- Should we 'spike' samples with a known reference compound or marker?
- Do time-honoured procedures in the literature, eg British Standards, or the British Pharmacopoeia, really have a 'constant' sampling error?
- Should flowing liquids be sampled in a laminar or turbulent condition?
- Is there a sedimentation gradient in the sample?
- Are there time-dependent variables of which we should be aware?
- Which statistical approach should be used to determine the number of samples to be collected?

There is undoubtedly a predisposition to take a 'manageable size of sample' and hope for the best. In the last resort the analyst may be forced into this situation, but if this is so, it is important to be aware that this is in fact what one is doing.

1.4.2. Pretreatment and Concentration

Once the field sample has been collected and transported to the laboratory, the analyst is usually faced with another group of activities prior to the actual acquisition of results. These may involve such activities as drying, measurement of weights or volumes, dissolution in solvents, separation from interfering materials, and preconcentration of the analyte.

Even drying a sample, which is often thought of as a trivial process, is not as straightforward as it seems. The water may be bulk water, (ie we have a simple wet sample), or it may be adsorbed on the surface or occluded in the structure, or even chemically combined as a hydrate. Furthermore, water is an inherent part of many biological samples and the sampling itself may affect the behaviour of plant and animal cells with respect to water. You can see that a well-developed protocol for drying is going to be important for some analyses. In some methods we dry samples to constant weight at a fixed temperature (often 110 °C) in others we dry them to the anhydrous state, in yet other methods samples may be 'air dried' or even analysed 'as received', followed by a separate determination of the water content.

Two of the most common activities in any analytical laboratory are weighing, and the measurement of volumes (or transference of known volumes). Weighing is almost always much more precise than volume measurement. Compare for yourself the error in say weighing one gram by using the conventional analytical balance found in almost any laboratory, with the error in volume measurement during a titration, by using pipettes and burettes. In the latter case you are lucky to secure better than $\pm 0.02\%$ in volume accuracy, whereas in weighing you expect to do better than $\pm 0.0001\%$ with no problems. In practice, of course, one may be trading off convenience against precision; it is quite common to use a combination of mass and volume measurements. In fact there are many circumstances in which the added precision is unnecessary, so you can see that methods should be evaluated with reference to the user's needs.

Even the simple, and apparently straightforward procedure of dissolving a sample should not take place outside the analyst's control. In particular with the accuracy of many modern instrumental techniques, solvent purity becomes important, even for water.

It is likely that any analytical samples you have already met in introductory courses will have been readily soluble in water. It is unfortunate that many real-life samples are not so readily soluble and the analyst encounters the partial dissolution of his samples. He is now faced with a decision between going for complete dissolution in increasingly 'aggressive' solvents or accepting a defined level of extraction. The latter is quite common in the chemical analysis of soils, partly because the siliceous material may be of little analytical interest, yet extremely difficult to get into solution.

Another process encountered in chemical analysis is *separation*. Essentially one is removing the material of interest from the field sample. There is a range of special techniques for this. These include ion-exchange, chromatography, solvent extraction, and electrodeposition; details of these are given in the later Units. The sort of questions we need to ask as part of our analytical plan are:

'Are the methods selective for a range of substances?' or

'Are they specific for one component?'

Naturally we shall be interested in interferences and in the possibility of cumulative errors. You will find later that techniques similar to those for separation are used for concentration of species to the levels accessible to our measuring devices.

1.4.3. Measurement of the Amount or Concentration of the Chosen Substances

This is of course the step which immediately springs to mind when one mentions analysis. We have already seen, however, that this is preceded by a variety of processes, all of which must receive proper consideration and control if the actual measurements are to be worthwhile. You will find that it is this part of 'analysis', ie the measurement itself, which accounts for the bulk of analytical texts and research papers on chemical analysis.

This part of the analytical plan is particularly concerned with knowledge of concepts drawn largely from physical chemistry. In other Units of ACOL you will see this continuous interlacing of physicochemical principles with the realities of acquisition of results. Typical theoretical areas which have to be developed within the analytical context are as follows:

— concentration, activity coefficients, equilibria, solubilities, redox behaviour, coordination models;

— spectroscopic behaviour, light absorption, natural line-widths, line-shape models, radiation quenching, dispersion;

— adsorption, desorption, polarity, ion mobilities.

In the various Units you will meet later in the series, it is important to recognise that the theoretical parts are there not simply for intellectual completeness, but to enable you to adopt an informed approach to laboratory practice. It is never satisfactory to simple stick a sample into a 'black box', twiddle the controls as required by the handbook and take a reading from a dial or VDU screen.

We now look briefly at the broad areas of particular interest to instrumental methods of analysis.

— **Presentation**

Presentation refers to the way in which the sample is exposed to the phenomena being used to supply a measurement. Having gone through the stages of collecting an ideally representative sample and perfect pretreatment, one can still meet artefacts of presentation. For example, some techniques utilise injection of the sample into a flame followed by observation of the flame's behaviour. Clearly, use of badly placed equipment subjected to draughts may invalidate hours of careful laboratory work and effectively waste the use of thousands of pounds-worth of dedicated instrument engineering. Similarly methods based on spark-excitation of a sample may vary, depending on whether the spark always goes to the same place or whether it jumps around on the sample surface. In some techniques such as infrared spectroscopy one often find highly sophisticated computer-controlled equipment, but the sample presentation depends on grinding materials in a mortar—a technique introduced by the ancient Egyptians!

— **Calculation**

In all instrumental techniques one is making a measurement of a signal which has undergone a variety of electronic and electrical processes. In addition, the generation of the signal itself may involve disruption of the analyte at an atomic or molecular level. Therefore, the analyst must be aware of the completeness of the theoretical model which describes these processes, as he converts an electronic measurement into an analytical concentration.

— **Evaluation**

The fundamental question here is 'What statistical significance is to be attached to the numerical output from the instruments, or indeed from classical wet chemical methods?' We are, of course, all familiar with the spurious appearance of precision generated by pocket calculators, which always fill the display with figures after the decimal point. With the incorporation of dedicated computers into instrumental analysis we must also be aware of the different significances of numbers, and be constantly asking when 'rounding off' is acceptable or indeed desirable.

Once we have the results, we need to ask a number of further questions in a rhetorical sense to round off our treatment of the plan.

— **Accuracy** – how does the result compare with the 'true' result?

— **Precision** – is our result internally self-consistent?

— **Errors** – are any errors determinate (systematic) or indeterminate (random), are they absolute or relative?

— **Deviations** – how do the results spread around mean values?

— **Uncertainty** – are there uncertainties with respect to the sampling, the chemical behaviour, the behaviour of the instruments, or the evaluation of utility to the user?

SAQ 1.4a

Mark the following as either true or false.

(*i*) The highly sensitive methods available to modern analysts have minimised the problem of complicated sampling procedures.
True / False

(*ii*) Modern high-precision techniques can be applied only to homogeneous samples.
True / False

(*iii*) Large sampling errors are not removed by high precision laboratory methods.
True / False

(*iv*) Sampling gradients should always be thoroughly investigated before laboratory determination.
True / False

(*v*) Weighing is always preferred to volume measurement because it is inherently more precise.
True / False

(*vi*) Wet samples should always be rendered anhydrous before measurement or determination.
True / False

1.4.4. Summary

In this introductory Part we set out to explore the philosophy of chemical analysis and the context within which analysis takes place. In our treatment of the steps in a chemical analysis (the analytical plan) we have identified some key 'markers' for development in the specialist Units. In the following few pages we deal with some short case-studies. These are included to illustrate the importance of the context of the analysis, ie its role within a framework set by other decision-making processes. You should use these cases to enhance your understanding of the role of analysts as full participants in the whole process, rather than the layman's view of them as the acquirers of results. In the case studies the chemistry itself is of secondary importance; it is subordinate to our main task of illustrating the context of analysis.

1.5. CASE STUDIES AND THE ANALYTICAL CONTEXT

Case 1

Several years ago an aircraft manufacturer encountered serious problems of air-frame failure. The material used was a zinc-based die-cast alloy containing 4.14% aluminium and 0.064% magnesium. This composition was essential for controlling the grain refining in the zinc, ie it prevented the formation of large crystals of zinc.

The alloy had been well tested and was known to be reliable as regards strength and fatigue resistance. Test-bed studies on a particular air-frame were all satisfactory, but production studies revealed metal fatigue and a series of hair-line cracks.

Many thousands of chemical analyses for aluminium and for magnesium were carried out on a wide range of samples. These were all found to be within the specification. More or less by chance, experienced analysts suspected the possibility of indium contributing to the problem. This was because indium was known to be present in the impurities collected from the retorts used in the manufacture of the zinc.

Analysis detected the presence of traces of indium in some but not all of the zinc samples used in the manufacture of the alloy, and further work revealed that minute quantities of indium led to the precipitation of magnesium at grain boundaries, so changing the alloy microstructure.

Comments

The message here in the general context of chemical analysis is the question:

'Is the method OK?' or

'are we looking at the wrong part of the problem?'

The methods for aluminium and magnesium were all well validated and could not be faulted. In this example, we see that when the method is correct but the product is still faulty, our question must be

'Are we addressing the right issue?'

This needs knowledge of the stages and processes before the use of the material and/or its analysis. That is, the analyst needs the 'up-stream' history of his samples.

Case 2

The problem of concrete structures showing signs of ageing by cracking several years after being cast, has been well known for many years. The presence of soluble sulphates in the 'fill' material has been identified as a significant contributor to this, and upper limits for soluble sulphates are defined in the relevant British Standards specification.

A material known as 'black ash' derived from mining-waste slag heaps had been subjected to analysis for soluble sulphate, and samples tended to be on the border of acceptability for use as a concrete fill. Some samples were within specification, but others were just outside it. However, when the material was roasted or 'dead burnt', it changed colour and became 'red fill'. Analysis of the red fill for soluble sulphate showed that it was within specification and satisfactory for use as a concrete fill. Nevertheless after several years some structures containing red fill displayed cracking and subsequent examination attributed the cause to an excess of soluble sulphate.

What can be done?

Π Consider each of the following options for a few minutes, before reading the comments that follow.

(*a*) Demand a tightening of the specification. This would of course have serious commercial implications as a new fill would have to be provided. Also the waste which would otherwise have been used would require disposal, or alternatively a cleaning process might be developed.

(*b*) Suspect the analytical method being used and develop alternative methods, preferably based on different chemical principles.

(*c*) Examine the current analytical method in use and check for possible interference; this may lead to method development.

(*d*) Consider the fate of all sulphur-containing materials in both red fill and black fill.

Comments on Options (*a*) to (*d*)

(*a*) Analysts are, of course, frequently working towards accept/reject criteria based on specifications similar to those in British Standards. These are usually set after exhaustive study by experienced practitioners. Although specifications are from time to time revised, the data in this case provide no justification for pursuing this. This sort of demand essentially says we are looking at the correct analyte and we are using an accurate and reproducible method so it must be something else which is wrong!

(*b*) Here we are in essence saying that we are looking at the correct analyte (the soluble sulphate) but the method is not up to the job. In practice several methods may be available for any particular analyte. A check by several different methods is always a good strategic move when things go wrong, unless the instrumentation required is not available, (a cross-check by another laboratory is also a potential ploy). In this particular case similar results were obtained by all methods used and by different laboratories.

(*c*) This is rather similar to option (*b*) and again the analyst is usually on the look-out for this sort of thing. The possibility of interference in the method may be checked by comparing methods which depend on different fundamental principles and/or by adding suspected interfering agents. However, if the interference occurs at the extraction or collection stage, this will not help.

(*d*) This approach considers the chemical context of the whole activity rather than the traditional limitations of laboratory analysis. Here this turns out to have been a fruitful exercise.

What actually happened?

As we noted above, approach (*d*) proved to be the key to the problem. An experienced analyst noticed the strongly characteristic smell of hydrogen sulphide during the extraction procedure for the red fill. This clearly represented a loss of sulphur from the sample. But why was black fill, which was within specification, satisfactory in use when similar red fill was not?

The explanation was as follows. When black fill is roasted some of the sulphate is in fact reduced by organic impurities of vegetable origin. This converts it into sulphide. The sulphide does not, of course, show up as soluble sulphate but nevertheless the sulphur is still present in the fill. In use there is slow oxidation so that the sulphide is slowly reconverted into sulphate.

The analytical methods had been well validated and the specification had been appropriately set. The context of the previous history of the material showed that the analyst had not been asked to determine the correct material. The term 'soluble sulphate' in this context was too restricted. These considerations show that 'soluble sulphate' plus 'oxidisable sulphur' (as sulphide) were the data which were commercially relevant.

Case 3

A pigment manufacturer makes a bright-yellow pigment based on lead chromate. The process starts to produce a dirty-brown product without warning.

What do we do?

Π Consider each of the following options for a few minutes.

(*a*) Determine the lead chromate in the dirty-brown material.

(*b*) Analyse the lead acetate and potassium dichromate of the starting materials.

(*c*) Advise that the plant manager checks production conditions such as stirring rate, temperature, pH, order of addition, etc.

(*d*) Check for impurities in the starting materials and in the brown product.

Comments on Options (*a*) to (*d*)

(*a*) It is most unlikely that the major component is anything other than lead chromate. Lead to chromium ratios might be marginally altered but the likelihood of detecting a change is slight. In fact, chemical and physical examination indicates that the product is identical with authentic yellow lead chromate.

(*b*) Again this is not a likely candidate. However, one must suspect the possibility of impurities, either present on delivery of the starting materials or getting in during pH control or from the reaction vessels. The problem is that the analyst simply cannot check for everything under the sun, but he might use the pH control and vessel/pipe-work metals as a starting point.

(*c*) This should be a routine part of an enquiry of this sort. In this particular case everything appears to be identical for both 'good' and 'bad' product batches.

(*d*) Again, we can see that this requires qualitative analysis before any trace methods can be applied. As in comments on (*b*) we need to know what to look for.

What actually happened?

The firm also supplied pigments to the pottery industry, where it is well known that particle size can have a profound effect on colour (colloidal gold producing purple glazes is a good example). In our case microscopic examination revealed a change in the particle size as the only observable difference between the brown and the yellow product. It looked as though we were seeking something which affects the rate of particle growth from the solutions, but which either stays in solution or is present at an ultra-trace level. Experienced analysts looked first towards other elements known to be present in lead ores and unrefined lead. Antimony was detected at trace levels in the lead acetate. Laboratory tests showed that although antimony was *not* carried over into the product at detectable levels its presence in the starting materials nevertheless affected the particles during precipitation, presumably by some nucleation process.

The identification of the target is crucial if the analyst is to respond usefully to an industrial problem.

Case 4

The Environmental Health Department Officers of a small town on the fringe of an area undergoing industrial decline, experienced a spate of complaints about foul smells from a small brook which skirts the edge of town. They were aware that upstream there were slag heaps of considerable age. These were used for small scale disposal of domestic organic waste: holes were bulldozed in the slag, these were filled with waste, and then covered over. The Local Authority was aware of the possible problem of leaching from these sites and had been measuring dissolved hydrogen sulphide in leachate for over ten years. This arose because organic acids from the waste were leaching sulphide from the slag. However no complaints had been received before, and rather puzzlingly the analytical record showed that levels of hydrogen sulphide had not changed significantly. The demand for 'analysis' made by local conservation groups left officials baffled, as they were under the impression that they were already having the necessary measurements done. However, local environmental analysts recalled other pollution-control problems related to the brook in question, which had since disappeared. These were related to a red cast in the water, which was tracked down to disposal of water, rich in suspended iron oxide, into the brook by a metal-finishing works. The works had since closed down. Analytical investigation revealed a dramatic decline in the amount of suspended iron oxide at the time of the foul-smell complaints. Laboratory tests showed that the presence of iron oxides had been instrumental in removing traces of hydrogen sulphide (presumably over many years). When the iron oxide discharge ceased then the hydrogen sulphide was released downstream.

Comment

This example illustrates again that the acquisition of the analytical data is only a part of the information required; the whole context is important for solving the problem.

Case 5

A manufacturer of plastic kettles carried out a number of pilot studies prior to launch of a kettle based on a phenol-formaldehyde resin. These resins are basically quite stable but some of the end groups on the chains can be hydrolysed to give formaldehyde (methanal).

Formaldehyde is a recognised health hazard. The pre-service studies all showed that formaldehyde release during simulated use was well within limits set by Environmental Health Authorities. However, after service for some months and extensive retailing by a large chain-store, the company was inundated with complaints of a 'bad taste' in use and the chain-store's own research analysts found high levels of formaldehyde in the water. This continued after prolonged use and could not be put down to an initial 'breaking-in' period. In desperation the company sent kettles to two academic institutions and a research association for further analysis. Unfortunately, one academic institution pronounced the formaldehyde release to be above the limits and one well below them; the research association found them hovering at the limit. The stage was set for a round of professional recriminations and the manufacturer was no wiser. He dare not put his kettles back on the market until the chain-store found them acceptable and he had no logical reason for selecting one set of results rather than another. The manipulative demands of the method were relatively slight, and could not explain the deviation experienced by the analysts.

When presented with the geographical locations of the various participating laboratories, one of the academic analysts noticed that they were grouped into hard-water and soft-water areas and felt that this might explain the results.

Analytical development work showed that it was not in fact the hardness itself which gave rise to the problem but the natural differences of pH of the local waters. Those areas with slightly acidic waters gave low formaldehyde release and those with slightly alkaline waters gave high release, even after thousands of hours of use.

The results were explained in terms of base-catalysed hydrolysis of the end groups. The manufacturer subsequently found that by changing the accelerator he could produce a plastic with greatly reduced numbers of terminal groups and in fact one which was suitable for use in alkaline water areas.

Comment

This case illustrates again the need for the full background to be available to the analysts. All the laboratories involved were competent and the method was perfectly valid. There was no doubt that the data were correct, but you can see that this was of no help to the manufacturer. His problem was not the analysis, it was what the analysis could tell him about his product.

1.6. ANALYTICAL METHODS

In this short Section we look briefly at the major methods of chemical analysis, divided according to the physical chemistry which underpins the methods. You will find the theoretical details relevant to analytical applications developed later in the specialist Units, alongside the treatment of the practicalities of the methods themselves. Our purpose in this review is to sketch out the underlying relationships between groups of methods, which might otherwise pass unnoticed, simply because one can meet them in isolation. The philosophical basis of all approaches is, of course, the belief that there are observable phenomena for which the intensity or magnitude of the observation can be related in a prescribed way (often rectilinearly) to the quantity of material present. In the concluding paragraph we shall consider briefly the question of uncertainty in our results, to see how this is to some extent an inescapable feature of nature.

1.6.1. Methods Based on Weighing

These are the *gravimetric methods* of classical analysis. Originally developed for a wide range of inorganic species, they can now be used for some organic materials. We should also include electrodeposition as a gravimetric method. For the former, our interest is in inorganic and organic precipitating reagents and a study of the chemical reactions between the reagents and our analytes. The appropriate parts of traditional physical chemistry here are equilibrium studies and solubilities. We usually want fairly large separable particles of pure material, so growth rates and occlusion, for example, will be of interest. For electrodeposition we shall be interested in discharge potentials in various solutions.

1.6.2. Methods Based on Volume Measurement

Again it is more than likely that you will have met this group of methods as classical volumetric analysis. The fundamental assumption here is that reactions have 'end-points', ie conditions for which the reaction is complete (and is, of course, quantitative). Typically one measures the volume of a reagent of known concentration which is equivalent to a solution of known volume but unknown concentration. The methods are usually grouped as follows: neutralisation (acid-base), complex-formation, precipitation, and redox methods. The underlying physical chemistry is drawn very strongly from equilibrium studies, and we need to know something about the relationship between concentration and activity. This group of techniques also requires the use of indicators, ie marker chemicals which tell us that some other reaction has reached completion. Here again chemical equilibria will be important.

1.6.3. Methods Based on Electromagnetic Radiation

There is a multiplicity of methods in modern analytical chemistry which exploit some aspect of the interaction of matter and radiation. In some ways we might see all of these as practical applications of quantum theory. Typically we treat the energies of atoms and molecules in terms of characteristic 'energy levels'. Matter can be raised to an energy higher than its normal or ground state by absorbing radiation; conversely when it returns to the ground state from some higher state it emits radiation. This gives us a convenient way of dividing our analytical methods into those based on *absorption* and those based on *emission* of radiation.

The absorption methods are usually classified according to the wavelength range of the radiation used (which means that each is dealing with a different physical process in the molecule). They are all forms of *spectroscopy*. The important analytical techniques are colorimetry, visible spectroscopy, infra-red spectroscopy (known along with Raman spectroscopy as vibrational spectroscopy), ultra-violet spectroscopy (a branch of electronic spectroscopy – colorimetry is also a branch of electronic spectroscopy), atomic absorption spectroscopy, nuclear magnetic resonance spectroscopy, and X-ray spectroscopy.

In all of these methods we are interested in the detection and measurement of the energy absorbed at a particular wavelength. Associated physical chemistry includes, for example, transition probabilities, non-radiative relaxation, and natural line-shapes. Many analysts regard it as a blessing that they do not have to delve too deeply into these highly theoretical areas.

The emission methods include flame photometry (a modernisation of the platinum-wire flame test), emission spectrography using a variety of excitation methods, X-ray fluorescence, and fluorimetry itself. In all these the fundamental process is the bombardment of the sample with excess energy (the excitation) to raise it to a high energy state followed by the observation of characteristic wavelengths of emitted radiation as it relaxes back to the ground state. The physical chemistry here involves radiation quenching, inter-system crossing, and upper-state populations, again mercifully of only peripheral interest to the analyst.

1.6.4. Methods Based on Electrical Properties

These methods will clearly all draw strongly on various aspects of electrochemistry; indeed one can view them as applications of electrochemical theory. The methods include pH techniques, voltammetry, polarography, potentiometry, and conductance methods. The exciting developments of ion-selective electrodes also fall into this class.

1.6.5. Other Methods

The remaining methods: mass spectrometry, thermal methods, radiochemical methods, and optical methods all form classes on their own without much in the way of shared basic theory.

You will be able to see from this very brief treatment, which is little more than a list of methods, that analytical chemistry draws upon a very wide range of fundamental physical processes. Some of the most powerful developments in modern analytical chemistry draw their strength from combining different methods, and year by year the range of applications is broadened and the limitations previously felt by analysts are reduced. Nevertheless you should always remember that in industry and commerce a vast number of chemical analyses do not require 'state-of-the-art' sophistication. The practising analyst will select his techniques with the principle of *fitness for purpose* clearly in his mind.

1.6.6. Analysis and Uncertainty

In the specialised section dealing with measurement and statistics you will deal with some of the ways in which one tries to remove uncertainty from the interpretation of our results. It is, however, worthwhile at this early stage in the programme to consider the different types of uncertainty.

'Real' Uncertainty

From a philosophical point of view we can recall Heisenberg's ideas about uncertainty. Without delving into the theory behind these ideas we can say that beyond a certain point, related phenomena cannot both be known with increasing precision. For the analyst this is important as it influences the width of the various 'lines' whose wavelength is measured in spectroscopy. The point is that instrument engineers may be able to narrow lines somewhat *but* there is a limit beyond which further improvement is impossible because of features fundamental to the natural world.

Chemical Uncertainty

This occurs because chemical systems are all potentially very complex. Luckily the complexity is restricted to a minute proportion of the material present. Equilibrium behaviour and features such as the relationship between activity and concentration contribute to this sort of uncertainty. The point of interest for the analyst is that the systems being studied are not to be regarded as fixed and immobile; there are always minute changes going on in all chemical systems and our interpretations of results must acknowledge this.

Instrumental Uncertainty

Instruments are, of course, real man-made units; radiation goes through slits, is reflected by mirrors or is focused by lenses, falls on photomultipliers; electrons flow through wires of different materials, etc. None of these is perfect in the sense of being absolutely without faults, and as a consequence there is always an element of uncertainty associated with the behaviour of instruments. Modern instrument manufacturers have, however, developed their skills to a level such that for the vast majority of applications instrumental uncertainty is negligible for well maintained and correctly operated instruments.

Psychological Uncertainty

Analytical chemists are still human! As such they are susceptible to the visual nature of the output from the various machines and instruments they use. Many of us have an irrational preference for seeing graphs rather than being presented with numbers, we feel more 'at home' with a spectroscopic trace than a computer print-out. However, traces based on the same data can end up looking different, for example a trace of absorbance, log absorbance or transmittance may produce a different reaction as the analyst makes his rough assessment of the results. There is no reason to believe that you will be immune from similar psychological aspects of uncertainty.

Summary

Analysis and Uncertainty

'Real' uncertainty	Fundamental uncertainty associated with all matter.
Chemical uncertainty	Uncertainty about the chemical behaviour of the sample.
Instrumental uncertainty	Uncertainty arising from the 'less than perfect' nature of man-made objects.
Psychological uncertainty	The analyst falls prey to subjectivity.

2. Introduction to Sampling

We can consider an analytical process to consist of six operational steps.

These are:

- Case history and identification of the problem,
- Choice of method,
- Sampling,
- Analytical procedure,
- Determination,
- Evaluation of the result and the report.

Of all these steps in the process, the taking of the sample is the most important. The sample eventually analysed will in most cases, be a very small part of the material to which the analytical result relates. And yet it is on the basis of this result that a commodity is, for example, purchased, sold or dispensed.

For example, let us consider a situation in which an ore is purchased, the price paid for the ore being determined by its metal content. If, because of incorrect sampling, the metal content of the sample analysed was found to be 5% w/w, when in fact the overall average figure was 4%, then your employers would be paying 25% more than necessary for this material.

As you will see later in this Unit, sampling can involve very complex procedures, often including many stages of sub-division before the analytical result is finally obtained. If, at any of these stages the sample taken is not *representative* of the bulk of the material from which it was taken, then the result reported will be inaccurate.

In modern terminology the general term 'sample' has been replaced by the more precise term 'representative sample'.

Definition of 'representative sample': *A portion of a material taken from a consignment and selected in such a way that it possesses the essential characteristics of the bulk.*

Now what do we mean by 'essential characteristics of the bulk'? The question is best answered by considering the sampling of a solid material having a variety of particle sizes. The sample will not be representative if any of these particles are excluded from being taken from the bulk because they are too large. Conversely, to choose large particles and not the dust also is not representative. One of the essential characteristics in this example therefore is particle size. The sample taken must be of the same composition as the material from which it originated.

We should, in fact, experience few sampling problems if all materials to be sampled were known to be homogeneous. Gases and liquids are often considered to be homogeneous though this is not always so. For instance, liquid flowing along a pipe flows faster in the centre of the pipe than at the wall. Thus some form of mixing must be considered before a strictly representative sample can be taken. Gases often separate into layers (stratification) due to differences in densities, and again some form of mixing may be required before sampling the gas. Solids, however, should never be considered as homogeneous until such time as they have been homogenised, normally by grinding them in a pestle and mortar, often made of agate.

Sampling, therefore can be a very complex procedure. Also, we have no way of knowing that the portion of the material eventually analysed does truly represent the bulk of material from which it was taken. Therefore, we rely to a large extent on a statistical approach to sampling, a reliable sampling procedure being developed for a particular situation only after considerable experimental work has been carried out. Even then the method actually chosen to take the sample is the one which *appears* to give the correct answer. The true answer, of course, is available only if all the material (population) is analysed or tested, obviously an impossible situation. Statistics therefore plays an important part in the design of any sampling procedure.

The most common and important source of error in the sampling of bulk materials is *bias*. Bias is defined as a systematic displacement of all observations in a sample from the true or accepted value, or a systematic and consistent error in test results. Bias is often caused by the human weakness of 'taking the easiest way out', or by simple lack of consideration. For instance, if we are asked to take samples for analysis from five boxes of confectionery in a room which contains fifty boxes, and we choose the five boxes that are closest to the door, then we have introduced bias into the sampling procedure. The sample cannot be statistically representative as we have failed even to consider the forty-five boxes furthest from the door. Bias can however be eliminated by the correct design of a sampling procedure.

Finally before we leave this introduction, let us consider briefly some important factors that must be considered when we attempt to develop an effective sampling procedure. These will be expanded in later sections of this Unit. They are:

- the sample taken must be representative of the bulk (population),

- the quantity of sample to be taken must be determined,

- the handling and subsequent storage of the sample must be correct.

SAQ 2a

Can you list some important factors that may need to be considered when deciding upon the sample size?

SAQ 2b

Why do you consider handling and/or storage of a sample may cause problems?

List of Objectives

When you have completed this section you will:

- be aware of the importance of the sampling stage in an analysis,
- appreciate the importance of taking a representative sample,
- be able to avoid bias within a sampling procedure.

3. Design of a Sampling Procedure

Definition of 'Sampling Procedure': *The succession of steps set out in a specification which ensures that the sample eventually taken for analysis shall possess the essential characteristics of the bulk (population).*

As we have indicated already, a sampling procedure can involve many stages before the analysis of the material is carried out. We cannot over-emphasise that as an analytical sample of only a few grams may relate to many kilograms of original material, great care must be taken at all sampling stages to ensure that a representative sample for analysis is taken. The various stages in which sampling may occur, and which may therefore form part of a sampling procedure are listed below,

- sampling unit,
- increment,
- gross sample,
- sub-sample,
- analysis sample.

The relationship between these stages is illustrated in Fig. 3a.

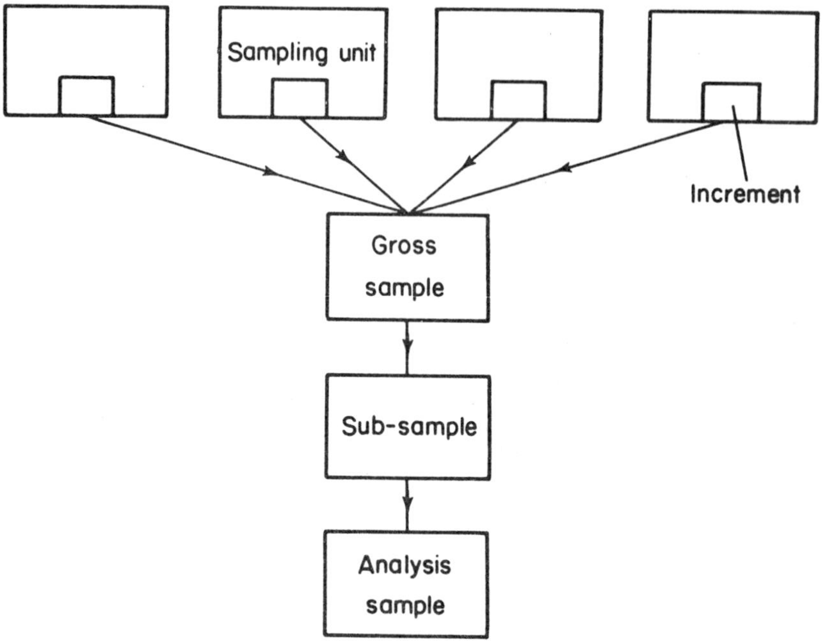

Fig. 3.1a. *Relationship of sampling stages*

A consignment of a material that requires sampling may consist of a number of discrete packaged units (eg drums, barrels, cartons). If each of these separate units or containers is to be sampled then each of these is called a *sampling unit*. The sample removed from each of these separate units is called an *increment* and the combination of increments a *gross sample*. The gross sample may well be too large to submit for analysis, in which case it requires sub-division in order to produce a *sub-sample*. The sub-sample supplied for analysis will be homogenised by the analyst before the *analysis sample* is eventually taken.

3.1. STATISTICS IN SAMPLING

It is not within the scope of this unit to consider in detail the statistics required to cover all aspects of sampling. However, it is impossible to discuss the principles of sampling without some recourse to statistics.

Statistical sampling is based upon the principle that all particles or portions of the material (population) should have an equal probability of being present in the sample taken. The parameter (eg the content of one constituent) which is determined during the analysis, is assumed also to have a normal distribution in the population. If it were possible to take an infinite number of samples from the population and to analyse for this parameter in each sample taken, then the frequency distribution of this parameter should obey the normal distribution law, which is illustrated in Fig. 3.1a.

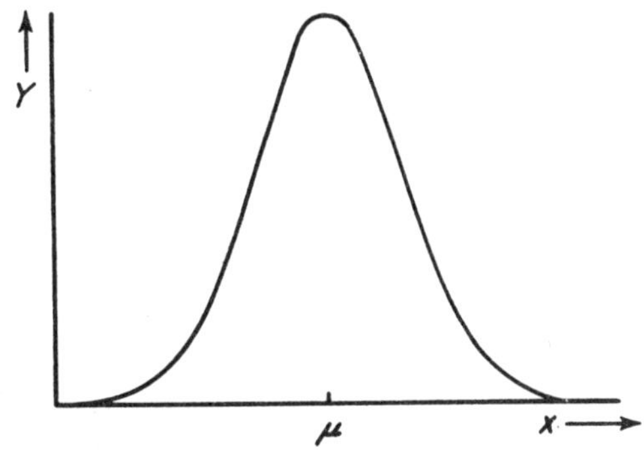

Fig. 3.1a. *Normal distribution curve*

The equation for the curve shown in Fig. 3.1a is Eq. 3.1,

$$Y = \frac{1}{\sigma\sqrt{2\pi}} \exp\left[-\frac{(x_2 - \mu)^2}{2\sigma^2}\right] \qquad (3.1)$$

where Y is the variance of the population,

σ is the standard deviation of the population,

μ is the mean value of the parameter,

x is an individual value of the parameter.

[Note: Variance is the square of the standard deviation].

The standard deviation is calculated by using Eq. 3.2,

$$\sigma = \left[\frac{\sum_{i=1}^{i=N}(x_i - \mu)^2}{N}\right]^{\frac{1}{2}} \tag{3.2}$$

where N is the number in the total population, and when the true mean of the result is known, or by using Eq. 3.3.

$$s = \left[\frac{\sum_{i=n}^{i=n}(x_i - \bar{x})^2}{n - 1}\right]^{\frac{1}{2}} \tag{3.3}$$

when μ is not known. In this case μ is replaced by the *estimated mean* $\bar{x}$, where $\bar{x}$ is the mean value of all the results obtained, σ is replaced by s where s is the sample standard deviation, and N by n, the number of results or observations. s gets closer to σ as the number of samples taken and analysed increases. However as the sample size decreases, the uncertainty introduced by using s to estimate σ increases. The extent of this uncertainty can be quantified by using Eq. 3.4,

$$\mu = \bar{x} \pm \frac{ts}{\sqrt{n}} \tag{3.4}$$

where t is a statistical term obtained from the '*Student's t-test table*' (see Fig. 3.1c) and n is the number of results from which the mean is calculated.

The value of $ts/\sqrt{n}$ is therefore the margin of error on either side of the calculated mean $\bar{x}$, within which we are confident that the true value lies. The extent to which we wish to be confident determines the value of t to be used.

We can relate standard deviation to the distribution law given in Eq. 3.1, and show the relationship in graphical terms in Fig. 3.1b.

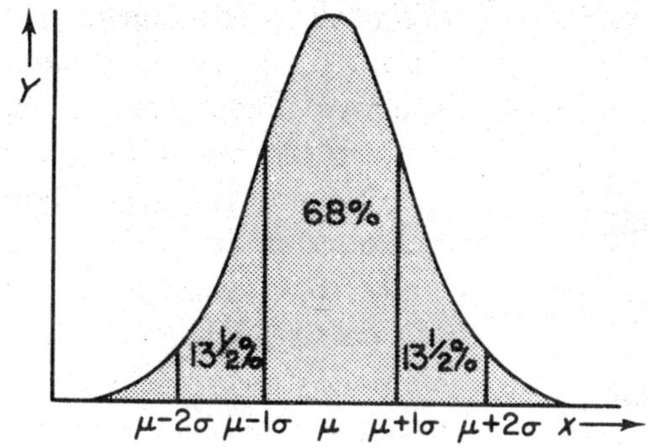

Fig. 3.1b. *Relationship of standard deviation to distribution*

From Fig. 3.1b we see that 68% of the analysis results obtained are within ± one standard deviation from the mean. About 95% are within ± two standard deviations.

In practice, of course, the individual constituents within the population generally do not have an equal probability of being present in the sample taken for analysis, because the sample will never contain only one completely uniform species. Hence there will always be an inherent variance (sampling error) in the composition of the sample eventually analysed. In addition to this, there will be the variance of the analytical result caused by the indeterminate (random) error in the analytical method adopted.

From Eq. 3.4 we see that the difference between the estimated mean value ($\bar{x}$) and the true mean value (μ) is given by the term $ts/n^{\frac{1}{2}}$.

If we give this difference the symbol 'E', then,

$$E = ts/n^{\frac{1}{2}}$$

By rearranging this equation:

$$n = [ts/E]^2 \qquad (3.5)$$

where s is the prior estimate of the standard deviation of the 'lot' of material to be sampled,

E is the maximum allowable difference between the estimate to be made from the sample and the actual value,

t is a probability value chosen so as to give a selected level of confidence that the difference is not greater than E. This can be obtained from the '*Student's t-test table*', Fig. 3.1c.

Degrees of freedom	Confidence level (%) 80	90	95	99	99.9
1	3.08	6.31	12.7	63.7	637
2	1.89	2.92	4.30	9.92	31.6
3	1.64	2.35	3.18	5.84	12.9
4	1.53	2.13	2.78	4.60	8.60
5	1.48	2.02	2.57	4.03	6.86
6	1.44	1.94	2.45	3.71	5.96
7	1.42	1.90	2.36	3.50	5.40
8	1.40	1.86	2.31	3.36	5.04
9	1.38	1.83	2.26	3.25	4.78
10	1.37	1.81	2.23	3.17	4.59
11	1.36	1.80	2.20	3.11	4.44
12	1.36	1.78	2.18	3.06	4.32
13	1.35	1.77	2.16	3.01	4.22
14	1.34	1.76	2.14	2.98	4.14
>15	1.29	1.64	1.96	2.58	3.29

Fig. 3.1c. *Values for t for various levels of probability*

In using this equation we must assume that the value of the standard deviation of the component in the lot has been determined. This generally requires not less than thirty determinations being carried out on a single lot. If this is so the value of t is equal to 1.96 which can be approximated by 2.0.

The overall variance is accepted to be the sum of the individual variances,

$$s^2 = s_1^2 + s_2^2 \tag{3.6}$$

where s_1 is the standard deviation of the sampling,

and s_2 is the standard deviation of the analytical measurement.

If one of the individual variances is more than ten times the other, the smaller of the two values will contribute very little to the overall value and for most practical purposes can be ignored.

It is important that we note at this stage, that the accuracy of a sampling procedure depends both upon the distribution of the replicate results obtained (precision) and upon the presence of any bias. It is, however, common in a sampling specification to assume that the prescribed method is unbiased, and thereby accuracy and precision become identical. If we assume that all the results obtained lie within a normal distribution pattern, then we can express precision as indicating those results which make up 95% of the statistical sample (ie which lie within approximately ± two standard deviations from the mean value). Thus, if we specify precision as being twice the standard deviation, then under unbiased conditions,

$$\text{Accuracy} = 2 \times (\text{standard deviation}).$$

At some point we must ask ourselves: 'How many analyses must be performed on a given population for the overall error to lie within some stipulated limits? 'Remember that analyses are expensive and often labour intensive, and thus both the customer and the analyst will want the number kept to the minimum needed to ensure the desired accuracy.

One of the main differences between the principles of sampling and of analytical determination is how we approach the question of error. The analyst is used to precise and generally reproducible techniques for performing an analysis, whereas the sample will often have to be taken from an extremely variable material over which the analyst has no control. The sampling error therefore is generally much larger than the analysis error. The sampling error cannot be eliminated entirely, only reduced, either by the use of improved sampling techniques or by taking larger samples. It is therefore necessary for us to be able to define maximum limits for sampling error, and to be able to work within these limits.

Π Let us suppose that as a result of making 30 determinations on a single lot of material, the value of the standard deviation for the component analysed was 0.187. If we set the maximum allowable difference between the estimate E and the actual value as 0.15, we can now calculate from Eq. 3.5 the approximate number of samples that we need to take.

From Eq. 3.5 we see that,

$$n = \left[\frac{ts}{E}\right]^2 \tag{3.5}$$

From the *Student's t-test table*, Fig. 3.1c, when the number of determinations exceeds 15, the value of t at the 95% confidence level is approximately equal to 2.

$$\therefore \quad n = \left[\frac{2 \times 0.187}{0.15}\right]^2$$

$$= 6 \text{ or } 7 \text{ samples}$$

Therefore, for samples of a similar type to that used to determine the standard deviation, we need to take seven samples in order to ensure the required sampling accuracy is attained.

SAQ 3.1a

Five samples are taken from a single lot of material, and on analysis are found to contain on average 5.20% of the component of interest. In previous experiments where lots of similar material have been sampled thirty or more times, the measured sampling variance has been calculated to be 0.16. Which of the following results should the analyst report for this analysis, given that he wishes to be 95% confident that his reported results are correct:

(*i*) between 4.70% and 5.70%,
(*ii*) between 5.06% and 5.34%,
(*iii*) between 4.85% and 5.55%?

3.2. DETERMINATION OF SAMPLE SIZE

The term 'sample size' has different meanings to the analyst and to the statistician. To the analyst, size refers to sample mass, sample volume, or sample dimensions, whereas to the statistician, the *sample size* refers to the number of separate units taken from a larger number of units (often termed a '*lot*'). In this Unit, the term sample size will refer strictly to the analyst's definition. All the necessary information about a material may not be available before the sample is taken. However, as experience is acquired about the material and, in particular, the distribution of components within the material, the sampling procedure can be adjusted to be more exact and more economic in meeting the requirements of the specification. *No sampling procedure can be properly devised without considerable previous experience of samples of the same type.*

If the material to be sampled is homogeneous, or can be homogenised readily, as for liquids and gases, then the whole lot may be considered as a single sampling unit, and a single increment maybe removed, the size of which is sufficient to provide for replicate analyses. However, with a heterogeneous material, it may be necessary to divide the bulk into a number of sampling units, one or more increments being taken from each unit. In Fig. 3a the increments are shown as being combined (composited) before analysis is carried out. However, there are instances where each increment is analysed separately.

The size of the sample to be taken from a bulk solid material depends upon;

— particle-size variation,

— homogeneity with respect to the component to be determined,

— degree of precision required in the analytical result.

Most solid materials consist of particles of various sizes and are often characterised by pronounced heterogeneity. Moreover, the species of interest maybe present in only a particular size or type of particle, though this occurs only rarely. If solid materials were made up from equally sized particles, with the constituent of interest uniformly distributed throughout them, the precision of sampling would be independent of the size of the samples. However, with real samples containing a variety of particle sizes, the minimum amount of the sample to be taken from solid materials depends upon;

— the size, number and shape of the granules,

— the average content of the species to be determined,

— the differences in specific gravity of the various components,

— the magnitude of the acceptable error.

Several attempts have been made to devise equations which allow the analyst to calculate the mass of sample to be taken from particular types of material. However, although these equations are based upon some statistical evidence, they are essentially empirical equations and contain constant terms which may vary considerably between batches of material. They should, therefore, be used only as an initial guide to sampling, the final decisions being made as a result of your own individual experience and results.

3.2.1. Composite *versus* Replicate Samples for Analysis

We have already shown in an earlier part of this section that a composite sample is produced by combining increments taken from individual sampling units, which together form a single lot. The results of analysing a sample produced by composition (combination) approximates to analysing each increment separately and averaging the analytical results. The method is mainly used when the variability of an analysed component within a lot is low, and so can be considered as insignificant. However, we must emphasise that different composite samples obtained from the same lot may yield different analytical results. This difference represents the sampling error of the lot.

The advantages of composite sampling are:

- less time is spent on analysis,
- sample variability is reduced or hidden,
- when carrying out trace analysis to detect the presence of contaminates, the identification of the contaminant in the composite sample will alert the chemist. Further investigation will then be carried out to find out whether the contaminate is distributed universally throughout the lot or is present only in a single sampling unit.

The disadvantage is that there is no indication of variation between sampling units.

When the variation between sampling units is expected to be large, the analysis of each sampling unit is recommended.

One of the problems of 'compositing' the various increments is that they are no longer available for individual analysis, should we detect a problem after composition. We should then have to return to the sampling stage and begin the whole sampling process again. A compromise procedure, therefore, is either to prepare the composite from only part of each increment, or to take two increments from each sampling point, only one of which is used to prepare the gross sample, as shown in Fig. 3a.

3.2.2. Identification of Lots

Problems sometimes occur when we attempt to sample a material which may consist of single or multiple lots. For packaged products, a lot is defined as *a collection of primary containers or units of the same size, type or style, which were produced under conditions as uniform as possible and are designated by a common container code or marking*. In the absence of any coding, a single day's production may also form a lot. Sampling as a single lot, material which in fact consists of multiple lots will inevitably increase the variability of the component throughout the matrix and thereby the sampling error.

The definition given above is satisfactory for separately packaged items such as processed foods, which are present in identical containers. The definition however is not satisfactory for bulk consignments such as shiploads of grain, or truckloads of potatoes. Here, even if primary identifiable containers are available such as sacks or compartments, repeated loading and unloading of the material will tend to disperse the lots. It is best to remove samples from bulk lots during the process of loading or unloading by using some form of continuous sampler.

SAQ 3.2a

The following terms relating to sampling have been used within this section:

- **analysis sample,**
- **gross sample,**
- **increment,**
- **lot,**
- **sampling unit,**
- **sub-sample,**
- **composite.**

Relate these to one another, and suggest at which sampling stage the maximum error is likely to occur.

3.3. TAKING OF INCREMENTS

Although we have considered that the identification of lots and sampling units may cause problems for the analyst, the major source of error in the overall sampling procedure will be the taking of increments, particularly of course, if the material is highly heterogeneous.

We have already shown that it is normal to combine all the errors inherent in a total analytical process and to express these as a variance (standard deviation squared). Therefore, if we let the combined variance (V) be equal to the sum of the individual variances,

$$V = V_s + V_a \tag{3.7}$$

where V_s is the variance of the sampling,
and V_a is the variance of the analysis,

then the accuracy (A) is given by Eq. 3.8,

$$A = 2(V_s + V_a)^{\frac{1}{2}} \tag{3.8}$$

or

$$A = 2V^{\frac{1}{2}} \tag{3.9}$$

We now have to consider what accuracy is required and how this accuracy can be achieved.

As we have seen in Eq. 3.8, the overall accuracy is a function of the variances of all of the stages participating in an analytical process, but which for convenience we have reduced to sampling and analysis only. Although not always so, usually the sampling variance will be significantly greater than the analytical variance. As a rule, we should attempt to minimise the analytical variance, and certainly not allow it to exceed about 20% of the total.

By rearranging Eq. 3.8 to give Eq. 3.10,

$$V_s = A^2/4 - V_a \tag{3.10}$$

we can calculate the sampling variance that we should be aiming to achieve, given the fact that we have defined the overall accuracy required for the whole of the sampling and analytical procedure. The achievement of this required accuracy now depends on the proper selection of increments together with the total number to be taken.

It is important to note that if periodic samples are being taken from the same source when there is considerable variation in the composition of the component with time, there is little point in attempting to achieve a highly accurate method of sampling. In these circumstances the increments would be bulked and an average figure found for the composite sample.

3.3.1. Placing of Increments

The increments are to be taken from the lot so as to provide an unbiased sample, with minimum random errors. Given that we have to sample stationary objects, then the sampling point(s) (places) must be chosen so that truly representative samples are obtained. For instance, when sampling a wagon-load of ore or coal, which during movement will have undergone a sieving process, samples must be taken that include all particle sizes. The positions from which these samples are to be taken are probably best chosen by use of random-number tables, the sampling unit having first been divided into separate imaginary parts.

Fig. 3.3a shows a single sampling-unit which has been divided into 24 imaginary sections: 8 at the top, 8 in the middle and 8 at the

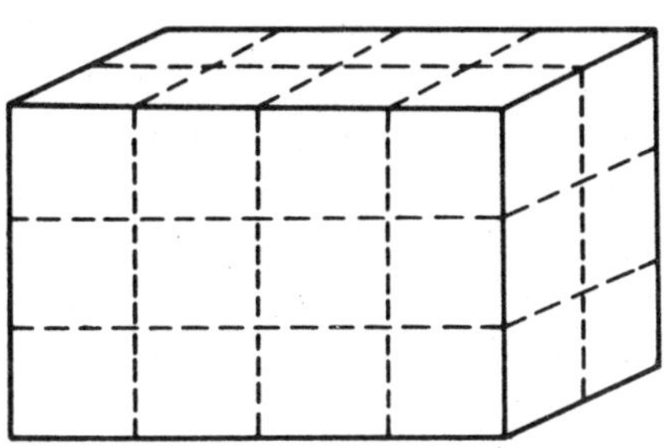

Fig. 3.3a. *Division of a single sampling unit*

bottom. If it is necessary to take samples from some of these sections, then each one is given a separate number, and the numbers of the sections to be sampled are obtained from a set of random-number tables. Assuming here that the increments are not to be analysed separately, then we must take care to ensure that the gross sample contains an equal quantity of material from each layer. This process is termed *imaginary sectioning*, and the number of sections into which the sampling unit should be divided is a function of the accuracy desired in the sampling.

With separately packaged units (eg cartons, drums), the use of random-number tables should avoid bias in the choice of the units to be sampled.

3.3.2. Size and Number of Increments

Bias in sampling may be caused in a number of ways. The main ones are:

— using an incorrect method of collecting the increment,

— an incorrect choice in 'placing' the increment

['Placing' the increment defines the point or part of the sampling unit from which an increment will be taken.]

As we have already emphasised, the major problem is likely to arise when sampling particulate matter of various particle sizes. Therefore, as a general rule, the size of the instrument used to collect samples should be approximately two and a half times the size of the largest particle in the material to be sampled.

Increments should generally be of the same size, except when a sample is to be removed from a flowing stream: the size of the sample taken should then be proportional to the flow rate of the material.

It can be shown from 'random sampling theory' that for material comprising particles of approximately equal sizes, the accuracy of the overall sample is determined by its total size, and not the size and number of the individual increments. This can be expressed mathematically as:

$$\text{Variance} \propto 1/\text{sample mass}$$

On the other hand, when the material consists of particles having a large range of sizes, then the number and size of the increments does affect the overall sampling accuracy. To reduce inaccuracy therefore, it is best to choose large-sized increments, selecting such a number that the required sampling accuracy is achieved.

If we express accuracy as variance, and assume that the component of interest is dispersed randomly throughout the lot, the overall sampling variance (V) becomes inversely proportional to the number of increments n (Eq. 3.11),

$$V = K_w/n \qquad (3.11)$$

where K_w depends upon the size of the individual increments and the variability of the lot. Given therefore that we have defined our overall sampling variance, and have some knowledge of K_w, then the number of increments, n, to be taken can be calculated. We must appreciate, however, that Eq. 3.11 is by no means exact, and provides only an estimate of n. The initial evaluation of K_w is accomplished experimentally, and is the average variance measured over a number of increments. Eq. 3.11 is generally considered to provide an over-estimate of the number of increments required, and should be modified when sufficient information is available concerning a particular system.

So far we have made no mention of the relationship between the size of the sample to be taken and the size of the lot. In theory the number of increments to be taken, and thus the total sample size, is strictly related to the variability of the component throughout the lot, and not to the size of the lot. In practice, of course, the larger the size of the lot, the greater the variability is likely to be.

3.3.3. Collecting the Sample

The taking of an increment may be only the first step in a sampling procedure prior to analysis. In Section 3.2.1, we compared the procedure of producing a composite sample from the increments taken with that of analysing each increment separately. If we aim to prepare a composite sample, then the increments taken are all placed directly into a single container of sufficient size to hold the total sample. However, if a further increment is taken from each sampling point, then these increments should be stored in separate containers for further investigation if required.

The size of the resultant gross sample maybe too large to send directly for analysis and therefore one or more sub-sampling routines will have to be used. Methods used for sub-division of gross samples will be considered later in this Unit (see Section 4.1.4). We should emphasise once again that the sample taken must be stored in such a way as to avoid a loss of constituents by evaporation or adsorption, or the contamination of the sample by the container in which it has been stored.

3.4. THE SAMPLING PROGRAMME

The way in which a sampling process is designed or developed depends critically upon the information that is required. Three important criteria used in determining a sampling programme are given below.

3.4.1. Sampling to a Specified Accuracy

It is often necessary for us to design a sampling programme to measure the average value of some property of a material, or to determine the average concentration of a component in a material. The procedure we must adopt is to take a number of samples and measure the property or determine the component in the samples taken. We can then calculate the variance between the results obtained. The number of samples n to be taken from future batches or during given periods of time is calculated by using Eq. 3.12:

$$n = 4V/A^2 \qquad (3.12)$$

where V is the variance between samples,

and A is the required accuracy of the mean.

[In order to calculate the variance initially, about thirty samples should be taken.]

SAQ 3.4a

Explain the relationship between Eq. 3.5

$$n = [ts/E]^2 \qquad (3.5)$$

and Eq. 3.12

$$n = 4V/A^2 \qquad (3.12)$$

where n is the number of samples taken,

t is the probability factor obtained from the t-test table,

E is the maximum allowable error,

V is the sampling variance,

A is the required sampling accuracy.

3.4.2. Sampling for Quality Control

Sampling for quality control was designed originally for the checking of mass-produced articles so that either the quality, quantity, or both were maintained. Most manufacturers supply their products to a given specification, and quality control charts are often prepared so that any movement or trend away from the specification can be seen and dealt with immediately.

A typical control chart is illustrated in Fig. 3.4a.

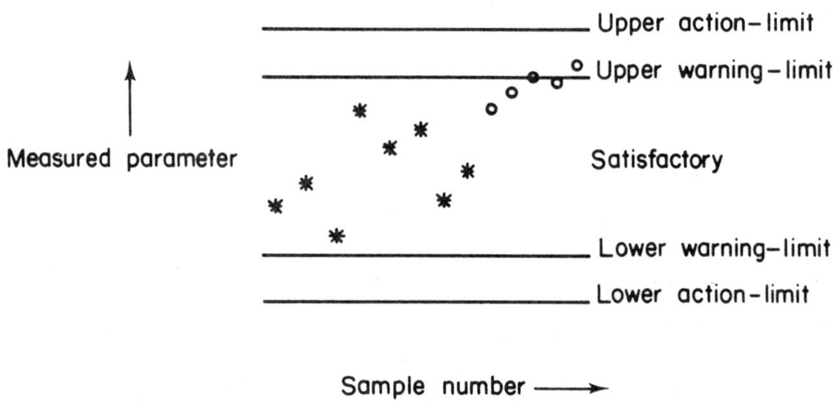

Fig. 3.4a. *Control Chart*

Samples are taken on a regular basis from the production line and under normal circumstances, the mean value of the measured parameter should be within the 'satisfactory' zone of the control chart (shown as * in Fig. 3.4a). If on the other hand, there is a trend towards the results appearing either high (shown as ○ in Fig. 3.4a), or low, then a warning is given that a fault is developing within the production process.

3.4.3. Sampling for a Decision

Sampling for a decision involves a programme of sampling which is continued to the point where sufficient information is available to make a decision. For instance, it may be necessary to sample a factory atmosphere, to check whether there has been a build-up to a dangerous level of a chemical normally employed in the production process. Samples would probably be taken from a number of points around the factory and the concentration of the chemical measured. If a dangerous concentration of the chemical is found to be present, then a decision should be taken either to shut down the process or to alter it sufficiently to reduce the atmospheric concentration.

Objectives

When you have completed this section you will:

- know of the various stages in an overall sampling process,
- appreciate the importance of statistics in sampling procedures,
- be able to decide where increments should be taken from, and how many,
- be aware of the reasons for sampling.

4. Methods of Taking Samples

In the two previous sections we have considered the principles of sampling. We now need to begin to consider just how we go about the process of taking samples, storing samples, sub-sampling, etc. Different techniques are required according to the state of matter to be sampled, and we must also take into account whether the material to be sampled is at rest or in motion. Before we begin to consider any of these points in detail, there are some general considerations that must be borne in mind.

Sampling Instruments and Containers

There is little point in our taking great care in extracting a sample, if it is liable to become contaminated, or a portion of it lost before the analysis carried out. For this reason therefore, sampling instruments and containers should be clean, and constructed of a material that is inert to the substance to be sampled. Also in order to eliminate the possibility of changes occurring in the sample, we should plan to analyse the sample as soon as possible after it has been taken.

Choice of Samplers

The choice of the 'right type of person' to carry out the sampling is not easy. The sampler must be conscientious, and capable of following precisely any given set of instructions. Above all, the sampler must be aware of the importance attached to the sampling procedure, and be capable of taking decisions when the unexpected occurs.

Automatic samplers are becoming increasingly popular, but they may be subject to bias unless they are rigorously tested before introduction.

Safety in Sampling

In certain circumstances, sampling can be an extremely hazardous operation. Safety harnesses and suitable protective clothing should be used when necessary, and often it is advisable that samplers should work in pairs for added protection. The sampler working alone and without adequate protection is likely to be less conscientious than is required for this important operation.

Reporting the Sampling

The sampler should report fully on how the sample was taken, indicating in particular any problems that arose. The sample taken should be clearly labelled with an unambiguous numbering system, and should include such information as:

- source of the sample,
- name of the sampler,
- date and time when the sample was taken,
- components to be determined.

The Dispatch and Storage of Samples

As we have already stressed, it is extremely important that the sample analysed has not been contaminated or transformed in any way as a result of handling, transport, or storage. Samples should therefore be held in air-tight metal, glass, or plastic containers capable of preserving the sample's integrity. Some common causes of the loss of sample integrity are:

- loss of volatile components,
- reactions of the components with air (eg with oxygen or carbon dioxide),
- decomposition in the presence of ultra-violet radiation,
- degradation caused by changes in temperature,
- changes due to catalytic activity.

A possible source of contamination which is often overlooked is that caused by the container material. Care must be taken to ensure that the container is either made from a material inert to the sample, or is coated on the inside with an inert layer (eg a metal drum having a plastic lining to avoid corrosive action by a stored liquid).

Most natural solid samples (eg ores, rocks) are very stable and can be stored and transported in various materials. Care must be taken, however, to ensure that no loss of fine particles occurs, and to this end the samples are often wrapped in plastic-coated paper or plastic sheet before being packed into wooden crates.

Some volatile materials are conveniently stored with the aid of a chemical reaction to form a less volatile product (eg hydrogen sulphide can be converted into cadmium or zinc sulphide). When no suitable reaction exists, the compound may be adsorbed onto an inert substrate (eg many organic compounds can be adsorbed onto activated charcoal, or modern propietary adsorbents such as 'Tenax'). In both these methods, the compound held can be recovered completely, when required for analysis.

SAQ 4a

Can you list some of the main considerations that should be borne in mind when taking a sample and submitting it for analysis?

SAQ 4a

4.1. SAMPLING OF SOLID MATERIALS

For the purpose of this section we shall consider solid materials to be either of a particulate nature or of a compact nature. Particulate material can be sampled in either a flowing or a static situation.

4.1.1. Particulate Matter in Motion

The flow properties of particulate matter are very much affected by particle size. Consequently, with materials consisting of particles of various sizes, it is essential that individual increments are large enough to allow even the largest particles a fair chance of being included in the sample.

Sampling of particulate matter (eg ores, coal) travelling along a conveyor belt can be achieved by removing a complete section across the belt. With manual sampling this is conveniently done by first stopping the belt and then removing all the material between two set points. The distance between these two set points is a function of the maximum particle size of the material to be sampled. As a rough guide, a width of up to three times the diameter of the largest particle should be adequate, when there is considerable variation in particle size.

Some automatic samplers are capable of removing a sample from a moving conveyor belt. The direction taken by the sampler should then be diagonal, in order to take into account the speed of movement of the belt, (Fig. 4.1a).

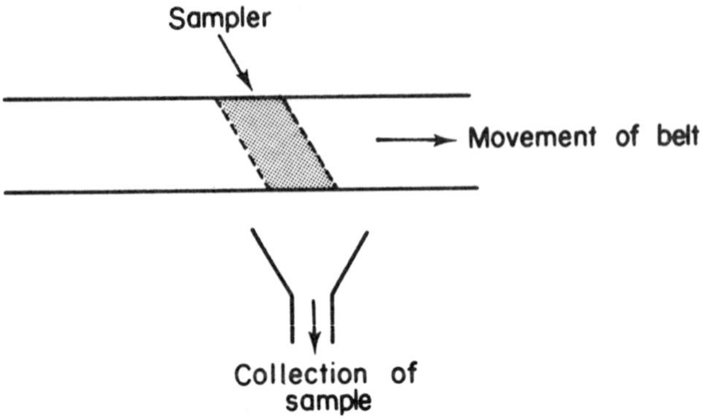

Fig. 4.1a. *Sampling from a moving conveyor belt*

The continual stopping of a conveyor belt to remove samples is somewhat disruptive, and it is often more convenient to remove samples from the end of the belt, where there is a free-fall situation. The sample will then often be taken automatically, by periodically transferring the flow of material to a sample container, as illustrated in Fig. 4.1b, or more simply by running a hard scoop across the free-falling solid.

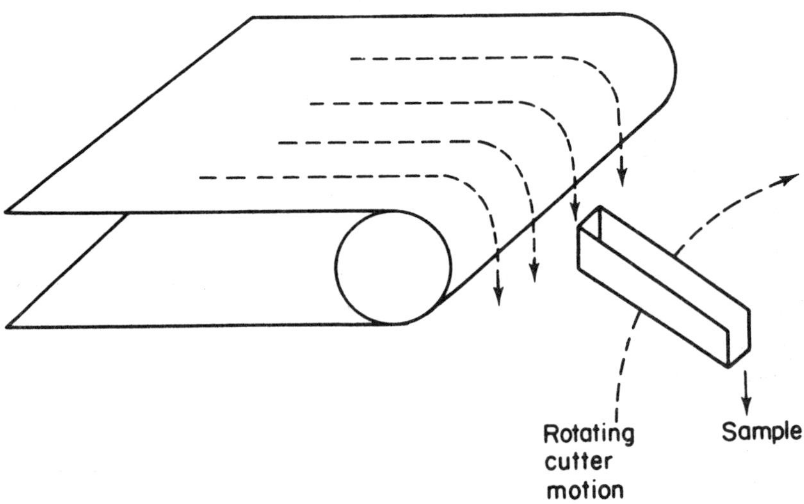

Fig. 4.1b. *Sampling from the end of conveyor belt*

When sampling particulate matter in a hopper, the size of particle is likely to be small and fairly uniform. A narrow sampling slot may thus be used, which traverses horizontally the end of the hopper, in order to take a representative sample of the whole of the cross-section present at the hopper outlet.

4.1.2. Particulate Matter at Rest

A fundamental condition for unbiased sampling is that the whole of the lot of material to be sampled is accessible to the sampler. Thus, the sampling of heaps, piles, wagons etc, is intrinsically unsatisfactory, and thus whenever possible, the material should be sampled in motion (ie on a conveyor belt or while loading or unloading). Sampling probes can be used to remove samples from the centre of a heap, but obviously it is difficult to ensure both the exact location of the increments, and that the increments removed contain the correct proportion of large and small particles. Less problems are encountered when the particles being sampled are of a similar size (eg those in sacks of flour, sugar, plastic granules)

The sampling probes used for such sampling operations are sometimes termed 'thiefs', and two such implements are described below.

(*i*) Bayonet or Split-tube Thief

The bayonet or split-tube thief is a tube with a slot running its entire length. One end of the tube is attached to a handle, the other end has a sharp cutting edge. The 'thief' is inserted into the container, with sufficient rotation so as to cut out a core of the material to be sampled. The 'thief' is then carefully withdrawn, with the slot uppermost, and the sample is removed from the slot. A typical example is illustrated in Fig. 4.1c (*i*).

(*ii*) Concentric-tube Thief

A concentric-tube thief which is suitable for sampling free-flowing particulate solids, consists of two tubes, one fitting tightly inside the other. The outer tube is pointed, and both tubes have holes cut through them at corresponding positions. The holes, which allow the sample to enter the thief, can be opened and closed by rotating the inner tube with respect to the outer tube. The thief is inserted into the material to be sampled, with the holes closed. When in position, the holes are opened and the sample taken. The holes are then closed again and the thief withdrawn. A typical example is shown in Fig. 4.1c (*ii*). A simpler example is also available consisting of a single outer tube into which holes have been cut. This can be inserted diagonally, with the holes uppermost, into the free-flowing solid, and carefully removed by an opposite action.

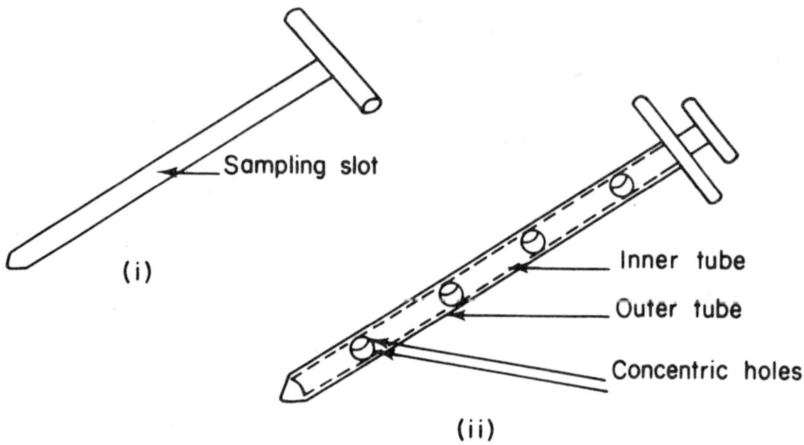

Fig. 4.1c. *(i) Bayonet or split-tube thief; (ii) concentric-tube thief*

4.1.3. Compact Solids

The term 'compact solid' is used to describe a variety of solid materials. These vary from natural materials (eg rocks and clays) to man-made materials (eg metal ingots, wire, concrete), and include compacted bales of fibrous material (eg wool and man-made fibres).

All the devices used for the sampling of compact solids use a cutting action to bore through the compacted mass. With augers (Fig. 4.1d) or split-tube samplers (Fig. 4.1c) the sample is held within the blades of the sampling device, to be collected on withdrawal. With drills, often working at high speeds, the sample is broken down to a small particle-size, and driven out of the hole created by the drill. All the drillings must be collected to secure a representative sample.

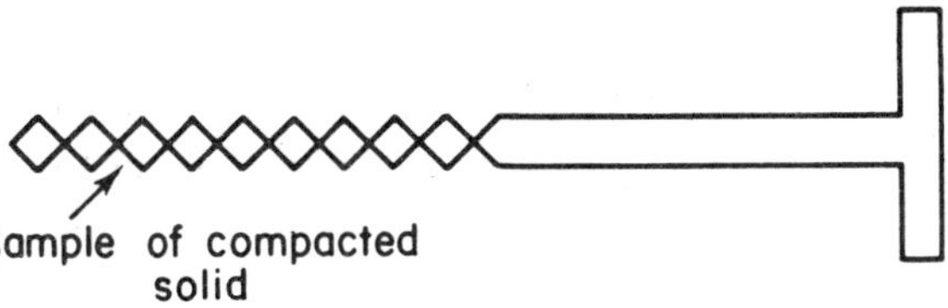

Fig. 4.1d. *An auger*

With finished metal products (eg wire, plate) the use of random drilling as a means of sampling either becomes impossible (eg wire) or unacceptable to the eventual sale of the product (eg plate). As these metal products are produced from batches of molten metal, which we assume to be essentially homogeneous, it follows that the finished product should be of an even composition throughout. Therefore, in order to avoid unneccessary damage to the product, samples for chemical analysis are normally taken at a convenient break in the production line, for instance, from the end of a reel of wire or of a sheet of metal. For wire or narrow metal sheet, we can take a cross section of the material as our sample. However, where the finished product is too large to make this feasible, then the careful filing of a sample from a convenient point is an acceptable alternative.

When the product is being sampled for a physical property (eg dimensions), we shall require a more random approach to sampling. But as tests of this type are on the whole non-destructive, they are unlikely to affect the value of the product.

4.1.4. Sub-sampling

Sampling of solid materials, particularly when there is a large variation in particle size, often produces a large quantity of material as the sample. Assuming that all of the increments have been 'composited', and that the analyst will not wish to receive these large amounts, it will be necessary for the gross sample to be sub-divided before submitting it for analysis. This normally involves a three stage process:

– crushing by using mechanical crushers, mills etc, until a mixture containing particles of the size required is obtained;

– homogenisation by using ball-mills,

– sub-division by using quartering or riffling techniques, which are described below.

Method of Coning and Quartering

The lot is placed on a flat surface, and by using a shovel, or other suitable tool according to the size of the sample, the lot is formed into the shape of a cone. Any fine material remaining is spread over the top of the cone. The top of the cone is then flattened, and the whole is divided into four approximately equal quarters. One pair of opposite quarters is then removed and formed into a separate cone for the process to be repeated. The procedure is illustrated diagrammatically in Fig. 4.1e.

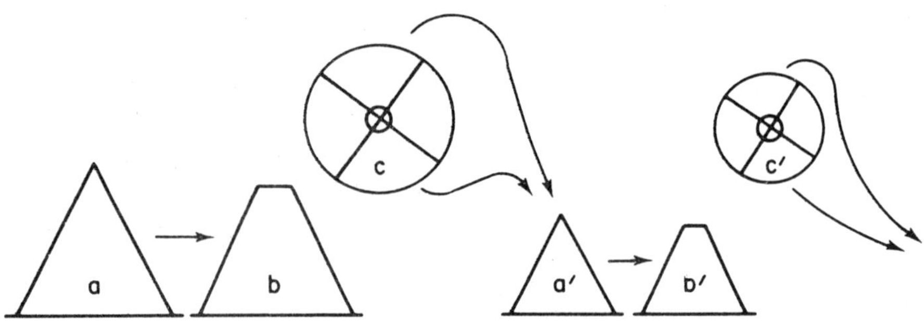

Fig. 4.1e. *Coning and quartering*

a and a′ represent the cone formed onto a flat surface,
b and b′ show the cone with the top flattened,
c and c′ give a top view of the flattened cones.

The Method of Riffling

A riffler is a mechanical device for dividing a sample into two approximately equal portions. A schematic diagram of a riffler is in Fig. 4.1f. The distances between the slots can vary and should be slightly greater than the size of the largest particle in the lot.

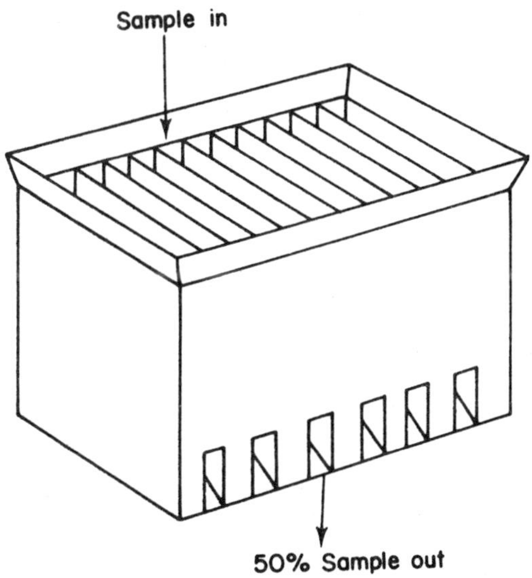

Fig. 4.1f. *Schematic diagram of a riffler*

The material to be sub-divided is poured into the top of the box and emerges on opposite sides in two approximately equal portions. As in the method of coning and quartering, the procedure is repeated a number of times until a sample is obtained which is of suitable size to submit for analysis.

Some modern rifflers, which are suitable for handling small quantities of materials (eg 1 to 2 litres), are based on a vibrating tray and rotating collection device. They are more efficient than the traditional riffler as they can sub-divide the sample into five or more portions in one procedure. A typical example is shown in Fig. 4.1g.

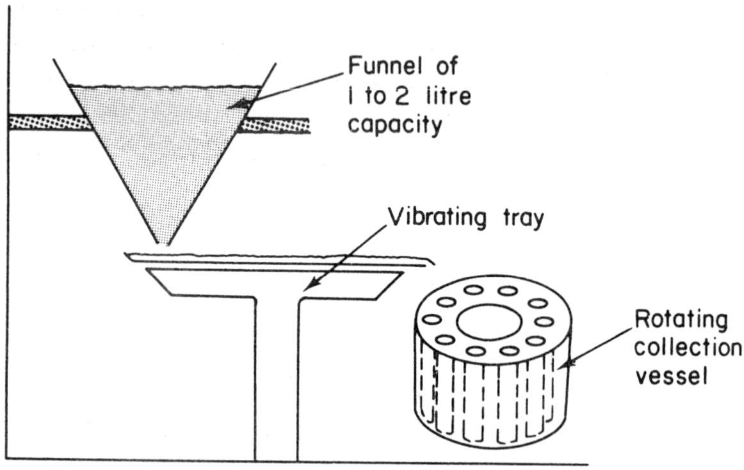

Fig. 4.1g. *A riffler suitable for handling quantities of material ≤2 litres*

SAQ 4.1a

In what ways may we take a sample from a conveyor belt?

SAQ 4.1b

> **What sort of equipment or method would you use to take samples from the following:**
>
> **(*i*) a dry sandy soil,**
> **(*ii*) a compacted sewage sludge,**
> **(*iii*) a 25 kg bag of smokeless fuel.**

4.2. SAMPLING OF LIQUIDS

At the outset, liquids appear to present less of a problem to the sampler than say, particulate solids. However this is true only when:

- the quantity of liquid to be sampled is small, and thus it can be homogenised by shaking it;

- the liquid is composed of only one phase.

Great care must be taken and attention paid when sampling larger quantities of liquid.

Liquids for sampling in bulk may be divided into four categories:

- those flowing in open systems (eg rivers, canals, industrial effluent),

- those flowing within closed systems (eg in pipelines),
- those stored in closed containers (eg in tanks, drums, carboys),
- liquids in open bodies (eg lakes, reservoirs).

We can now consider each of these situations in turn.

4.2.1. Liquids Flowing in Open Systems

The chemical composition of a flowing water may vary according to changes in a number of parameters (eg temperature, flow rate, distance from source, depth, pollution, sources), none of which can be controlled by the sampler. Because of this large number of factors, it is difficult to draw accurate conclusions as a result of a single sample being taken. The full information may be available only after a large number of samples have been analysed.

The samples are commonly collected by using wide-necked bottles or canisters, which are immersed at a suitable point in the flowing stream. For sampling at various depths, a weighted glass bottle with a removable stopper is a simple but effective device. A typical example is illustrated in Fig. 4.2a.

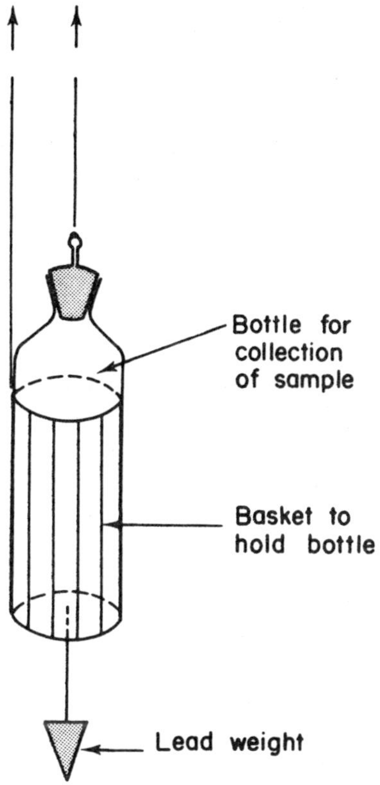

Fig. 4.2a. *Sampling device for liquids in an open system*

From the surface of the lake, say, the bottle is lowered into the water to the required depth, as measured by the length of rope or chain holding the basket. The stopper is then removed and the sample taken. The device is then hauled back to the surface and the sample is transferred to a suitable container for storage.

4.2.2. Liquids Flowing within Closed Systems

When we sample liquids flowing within closed systems, it is important that we first take into account the flow-rate of the liquid. At low flow-rates laminar flow predominates. This results in the liquid flowing at a velocity maximal at the centre of the pipe and decreasing to zero at the wall. In order to ensure homogeneity. therefore, it is preferable that we create turbulent flow just before the point where the sample is taken. Fig. 4.2b shows two ways in which we can create turbulence. Liquids flowing at higher rates already flow turbulently, and thus we need take no further action before the sample is taken.

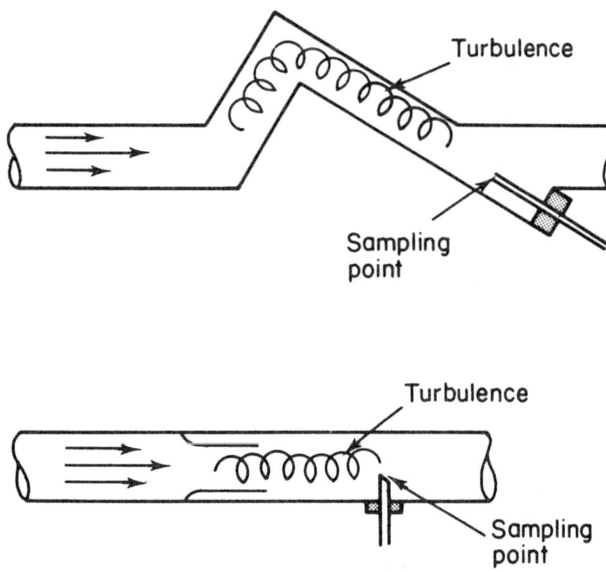

Fig. 4.2b. *Sampling device for liquids flowing within closed systems*

Note that the sample is removed, where possible, from the direction opposite to that of the flow. This is to avoid the sample being forced out of the sampling tube by the flow of the liquid. The diameter of the sampling nozzle is chosen to ensure the correct sample flow-rate into either the collection vessel or the analyser.

4.2.3. Liquids Stored in Closed Containers

Any liquid stored in a tank is liable to stratification due to the differing densities of liquids held within the tank. Therefore, if it is impossible to homogenise the lot prior to sampling it, we must take increments at various depths. Devices are available which enable us to carry out this task, and a typical example is illustrated in Fig. 4.2c.

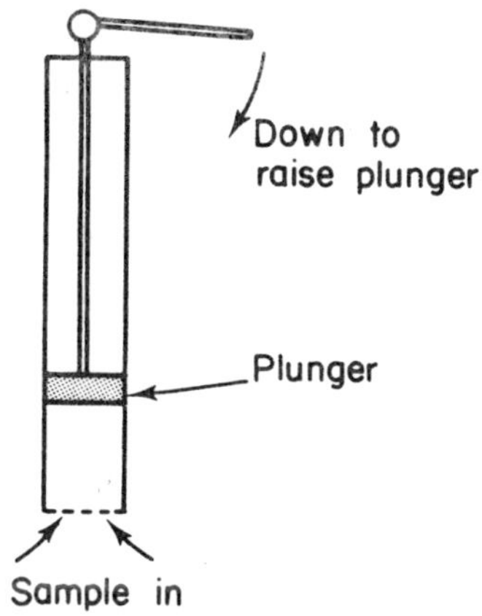

Fig. 4.2c. *Sampling device for an inhomogeneous liquid*

The device is lowered to the depth required. By raising the plunger inside the sampling tube, the liquid is drawn into it from the depth to which the tube has been lowered. The device can then be removed, the sample transferred to a suitable container, and the sampling repeated at a different depth.

The alternative procedure, which is often favoured for smaller containers (eg drums), is to sample the complete depth of the liquid. This is achieved by using an open-tube sampler, that is lowered carefully into the liquid and closed mechanically from above before it is removed. A typical device is illustrated in Fig. 4.2d.

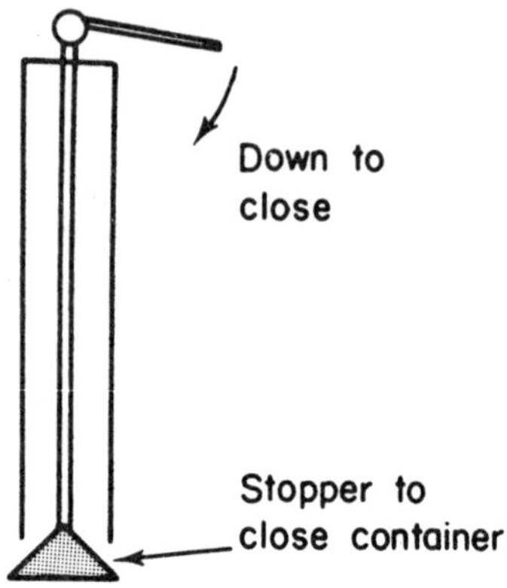

Fig. 4.2d. *Sampling device for liquids in small containers*

Once the sample has been received by the laboratory, any further sub-sampling can be carried out after shaking the sample to ensure that it is homogeneous.

4.2.4. Liquids in Open Bodies

'Open bodies of liquids' signifies large volumes of liquids in essentially static situations. A lake is a good example. In order to obtain a representative sample, it will be necessary to sample the liquid at various depths, and this can be done by using the weighted glass-bottle illustrated in Fig. 4.2a. However, if the lake is not too deep, and the sampling is to be carried out on a regular basis, then it is worthwhile constructing a permanent sampling device as illustrated in Fig. 4.2e.

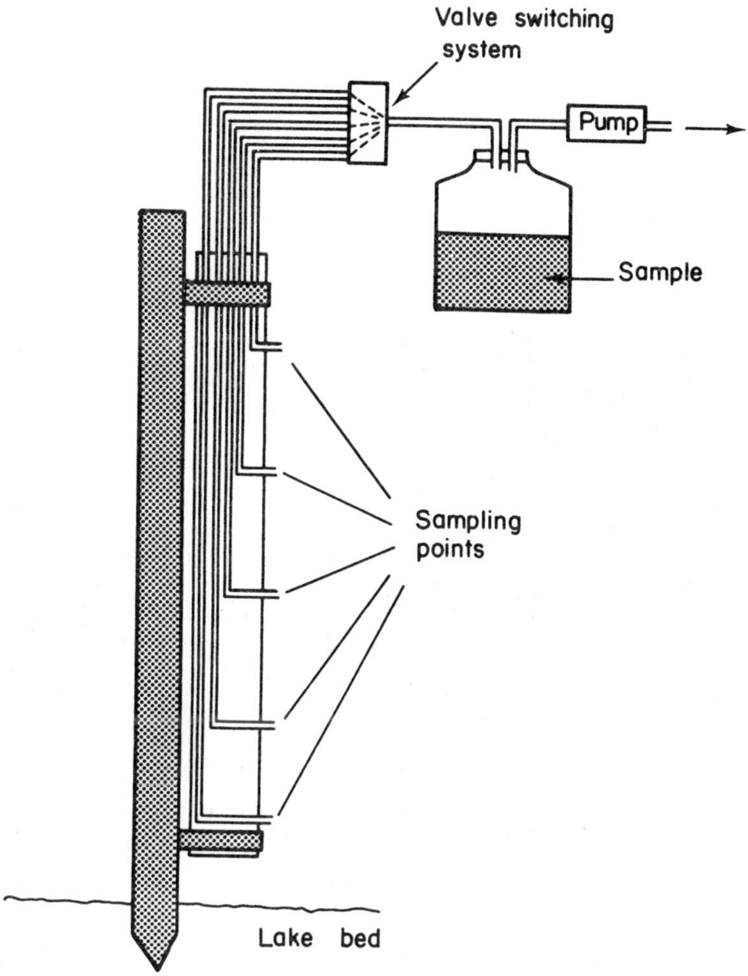

Fig. 4.2e. *A permanent sampling device for open body of liquid*

The device is constructed from inert plastic materials and allows increments to be taken at various depths, to produce either separate or composite samples. We must emphasise, however, that a device of this type is useful only for shallow water.

SAQ 4.2a

> **What precautions must we take when sampling liquids that are flowing slowly?**

SAQ 4.2b

> **What differences in sampling procedure would you adopt if you were required to sample a liquid of low viscosity inside two drums, one of which was of 10 dm^3 capacity and the other of 200 dm^3 capacity.**

4.3. SAMPLING OF GASES

As for solids and liquids, we must consider the sampling of gases in both a static and a flowing situation. Also we have to contend with an additional problem caused by a fundamental property of gases. This is, that the mass of gas present in a given volume, is a function of its temperature and pressure, and thus the composition of a gas sample may vary as a result of a change in either of these parameters. Therefore, whereas, it is relatively easy to store solid and liquid samples without significant losses of components, the storage of gases may present difficulties when a sample is taken at a temperature and/or pressure that is higher than ambient. This problem is highlighted in process analysis, when gases containing significant quantities of water vapour are sampled at high temperatures. In order to avoid storage problems, the direct transfer of the sample to the analyser is the recommended procedure.

4.3.1. Gases in Motion

The sampling of moving gases most frequently takes place from gas mains or in chemical plants where gases either form part of the feedstock or are one of the products. Because of the high rates of diffusion of gases and the high flow rates at which they are usually moving, the flows can be considered turbulent, and the gases homogeneous with respect to cross-section. Whether they are also homogeneous in time depends upon the nature of the chemical process and the opportunities available for gas-mixing before the sample is taken.

As illustrated in Fig. 4.2b, the sampling probe may be connected permantly into the pipework, enabling samples to be taken either randomly or at pre-set intervals. The samples are transferred *via* a delivery line either into the analyser directly or into a suitable collection vessel. The delivery lines are often heated to avoid changes in the composition of the sample before it reaches the analyser or the collection vessel. If the pressure of the gas being sampled does not significantly exceed atmospheric pressure, then we shall have to remove the sample by using a suitable pump.

The sample-collection vessels vary in size between 250 cm^3 and 100 dm^3, and are generally constructed from glass or a suitable metal which is inert towards the components in the gas-sample taken. You should note that steel vessels are not suitable for the storage of gases containing carbon monoxide, because of the formation of iron carbonyls. For the sampling of gases at very high pressures, suitably constructed pressure vessels are required.

4.3.2. Atmospheric Sampling

Atmospheric sampling is carried out to monitor levels of pollution in the air, and for convenience, we can divide sampling situations roughly into three categories:

- narrowly defined areas (eg boreholes, chimney stacks),
- large enclosed areas (eg factory atmospheres),
- open atmospheres.

We need to sub-divide atmospheric sampling into these three categories because any atmospheric sampling device will collect only the sample at the inlet to the sampling apparatus. Single samples taken from narrowly defined areas are generally representative with respect to cross section, although not necessarily to time. However, over a large enclosed area, we cannot assume the atmosphere to be of the same composition throughout the enclosure, as the sources of pollution will tend to be specifically located. A single sample is therefore most unlikely to be representative of the whole atmosphere. Further problems, which are encountered in the sampling of open atmospheres, are caused by natural phenomena (eg wind, precipitation, convection currents), and once again, a single sample is unlikely to be representative.

Sampling procedures to overcome these problems therefore involve either:

(*a*) simultaneous sampling at a number of separate locations over the areas being covered, samples being taken either randomly or at fixed time-intervals;

(*b*) continuous sampling over a long time (eg 12 or 24 hours) at a single location (this includes personal monitors);

(*c*) discontinuous sampling over a long time to produce a composite sample.

Procedure (*a*), although the most time-consuming, does provide us with information regarding the sources of pollution together with changes in pollution levels over given times (eg a normal work day). Atmospheric monitoring of this magnitude frequently involves autosampling followed by direct autoanalysis.

Procedures (*b*) and (*c*) are often used when sudden high levels of the pollutant are not expected, but continual exposure to the compound may have long-lasting effects (eg exposure to sulphur dioxide). In both these procedures, a composite sample may result from sampling during a given time period, or as in procedure (*a*) above, the samples may be taken and analysed automatically.

If we are collecting a single composite sample over a fixed time period, then in order not to have a large volume of gas to manipulate, the compound of interest maybe trapped, either chemically or physically, at the time the sample is taken.

Chemical trapping may involve, for example, precipitation, or complex formation to produce a coloured product, whilst physical trapping often involves the adsorption of the substances onto an inert substrate. The adsorption occurs as a result of:

(*a*) drawing a gas-sample through a tube which is packed with a suitable adsorbent (eg activated charcoal or a polymeric adsorbent);

(*b*) allowing a gaseous substance to diffuse through a thin membrane into a tube packed with an adsorbent (eg 'Tenax' – 2,6-diphenyl-*p*-phenylene oxide)

Both these adsorption methods are suitable for the sampling of air containing trace levels of organic components (eg solvent vapours). In adsorbers of type (*a*) the adsorbed substances are removed by solvent extraction before analysis by chromatographic methods. In type (*b*) devices, the adsorbed substances are generally removed by thermal desorption (controlled heating of the tube), directly into a gas liquid chromatograph.

Samplers of the type (*b*) are often termed passive samplers, and are useful as personal monitors, to be worn by scientists and technicians working in potentially hazardous enviroments. They have an advantage over other methods, in that they require no pumps to draw the sample through the adsorbent, and can therefore be worn without discomfort for long times.

Π Can you suggest a possible disadvantage of the use of passive samplers as personal monitors?

The main disadvantage of passive samplers is that they provide information of essentially historical interest, and cannot warn the wearer of the presence of sudden high levels of pollutants. Analysis of the tube samples is generally carried out within 24 hours of the sample being taken, so that some form of remedial action can be taken, if the wearer has been shown to be exposed to high levels of pollution over the period that the sampler was worn. Passive samplers are often used as a back-up to more sophisticated automatic sampling and analysis techniques.

4.3.3. Methods of Taking and Storing Gas Samples

As indicated above, gases should be analysed as soon as possible after being sampled. There are however, many circumstances where this is not feasible, and samples of gas will have to be taken, for analysis in your laboratory at a later stage. Samples of gas can be taken and stored in a number of ways, some of which are listed below:

— aspiration through glass vessels,

— displacement of liquid and storage over liquid,

— expansion into an evacuated vessel,

— by use of a sampling pump.

Let us consider briefly, each of these in turn.

(*a*) Aspiration Through Glass Vessels

The gas sample is pumped through one or more glass cylinders which are fitted with suitable stopcocks at either end. By closing the stopcocks, a sample of the gas is trapped inside the cylinder. Metal stopcocks that do not require greasing are recommended for these vessels.

(*b*) Displacement of Liquid and Storage Over Liquid

A gas pipette or another suitable container is filled with liquid in which the components are insoluble. One end of the pipette is connected to a reservoir of the same liquid by means of a piece of flexible tubing. A typical apparatus is shown in Fig. 4.3a. Liquids which are suitable for this purposes are mercury, water, salt solutions, and acid solutions. Mercury can be universally applied, but all of the others have limited application.

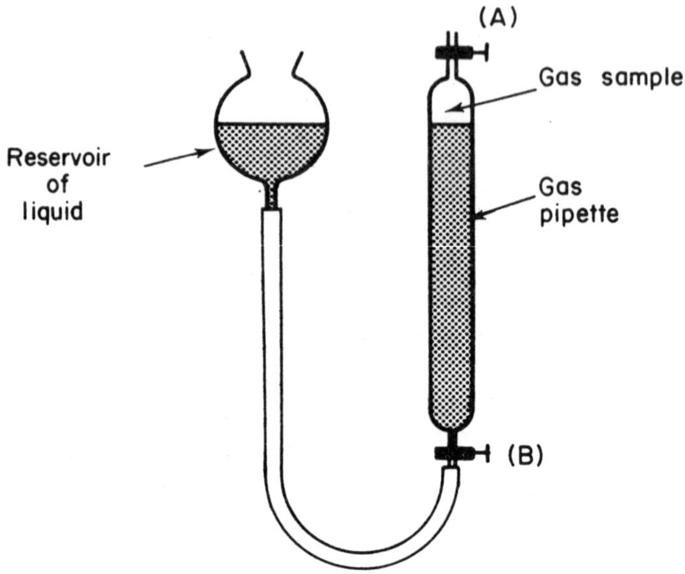

Fig. 4.3a. *Gas sampling by displacement of liquid*

By lowering the height of the liquid reservoir and opening the taps (A) and (B) of the gas pipette, a sample of the gas present at the point (A) will be drawn into the pipette bulb. When sufficient sample has been taken, the two taps are closed, and the apparatus can be transferred to the laboratory for analysis of the gas sample. It is more convenient however when the sample has been taken to close tap (A), then to raise the height of the liquid reservoir, to pressurise the gas held in the container. Tap (B) is now closed and the gas pipette is disconnected from the reservoir and transferred to the laboratory.

(*c*) Expansion Into an Evacuated Vessel

A suitable container, a gas pipette or similar vessel, is evacuated within the laboratory and taken to the sampling point. The sample is allowed to enter the vessel until atmospheric pressure is reached. The vessel is then sealed and returned to the laboratory for analysis.

(*d*) Sampling Pump

This is a hand-held device, approximately the size of a bicycle pump, and works in a roughly similar fashion. A typical pump is illustrated in Fig. 4.3b.

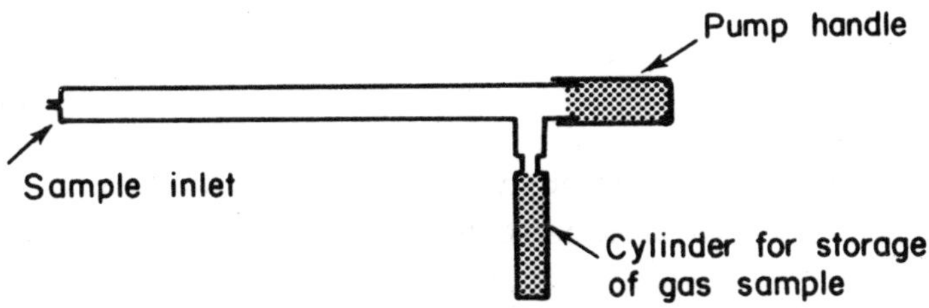

Fig. 4.3b. *Pump for gas sampling*

A gas sample is drawn into the pump on the backward stroke, and forced into the small cylinder attached to the pump, on the forward stroke. When sufficient sample has been collected, the cylinder holding the gas is disconnected and returned to the laboratory for analysis. To sample the gas inside the cylinder, it is connected to a pressure-release valve, and analysis samples are taken at various stages of discharge. This latter procedure is particularly important when the sample for analysis is thought to contain a mixture of permanent gases and low-boiling-point liquids.

Π Why may low-boiling-point liquids cause problems, in the sampling of gases which are held under significant positive pressures?

The boiling-point of a liquid is increased with increased pressure. Thus compounds that are gases at room temperature, but are not permanent gases, may become liquids under high positive pressures (eg in cylinders of butane and propane). If inside a pressurised gas cylinder, we have a mixture of compounds, only the vapour being sampled, then the composition of the vapour will not represent the true composition of the contents of the cylinder. Initially, the vapour will be richer in the more volatile components, to be replaced by the less volatile components as the cylinder is emptied.

SAQ 4.3a

Why are gaseous samples, more likely to suffer a loss of sample integrity on storage, than either solid or liquid samples?

SAQ 4.3b

In the text we have divided atmospheric sampling locations into three categories;

— narrowly defined areas,
— large enclosed areas,
— open atmospheres.

In what way are these three divisions different as locations to be sampled?

In answering this question you should take into account the representative character of the sample both with respect to cross-section and to time, and whether the atmosphere was static or moving.

SAQ 4.3b

4.4. METHODS FOR OBTAINING THE ANALYSIS SAMPLE

The quantity of sample presented to the analyst should always be greater than that required for a single determination, in order to allow for replicate analyses. The analyst must therefore be prepared to sub-divide representatively the sample provided, before performing the analysis. Let us consider how this can be achieved for solid, liquid or gaseous materials.

4.4.1. Solid Samples

Assuming that the sample presented for analysis is dry, but too large for it all to be ground in a pestle and mortar, the size will have to be reduced until a sub-sample is obtained which is small enough for us to handle (eg 10 to 100 g). The method of coning and quartering, and of riffling, as described in Section 4.1.4, can also be used in this context, to produce representatively, the sample-size required. The resultant sample is then ground in a pestle and mortar (agate) and finally sieved to pass through a 10^4 mesh/cm^2 sieve. The analysis sample is taken from that material which passes through the sieve. The sieved sample can now be stored in a screw-cap bottle and labelled 'sample as received', or dried in an oven at 105 to 110 °C for 1 to 2 hours, and then stored in a desiccator.

With samples that arrive 'wet', for instance clays or soils, the sample must first be dried before we proceed with the sub-sampling routine.

4.4.2. Liquid Samples

Single-phase liquids of low viscosity should present the analyst with few sampling problems. The sample as received is shaken to mix it, and an aliquot of it is removed by pipette, syringe, or by pouring. If the liquid is too viscous to be mixed effectively by shaking it, then it can be stirred by using a paddle stirrer to ensure homogeneity before the sample is taken.

Liquid samples which appear to consist of two phases, can be treated in one of two ways:

- by adding a suitable emulsifier to the sample and shaking it to produce an emulsion,
- by measuring the ratio of the two phases in the sample supplied, and then sampling each of them separately.

Of the two methods, the latter is to be preferred, as emulsions once created, can be extremely difficult to break down in order to separate the components.

4.4.3. Gaseous Samples

Once a gas-sample has been physically homogenised, or has achieved homogeneity by natural diffusion, it will retain this equilibrium composition, provided it is not subjected to any physical or chemical change. The sample received by the laboratory will be present in either a glass or a metal cylinder, or possibly in a plastic bag, and should be at or above atmospheric pressure.

The analysis samples are generally taken from the sample received, by using gas syringes or constant-volume sampling loops. Although syringes may be used to sample gases which have been stored at or about atmospheric pressure, the loop type sampler may be used only when the pressure of the gas is significantly higher than atmospheric.

To remove a gas sample from a gas-pipette or similar cylindrical holder, we can attach a small piece of silicone rubber tubing to the tap at one end of the holder and crimp the tubing tightly with a screw clip. The tap or stopcock can now be opened and a syringe with a long needle attached, pushed carefully through the silicone rubber into the cylinder. A sample of the gas held in the cylinder may then be withdrawn into the syringe. The whole device is illustrated in Fig. 4.4a

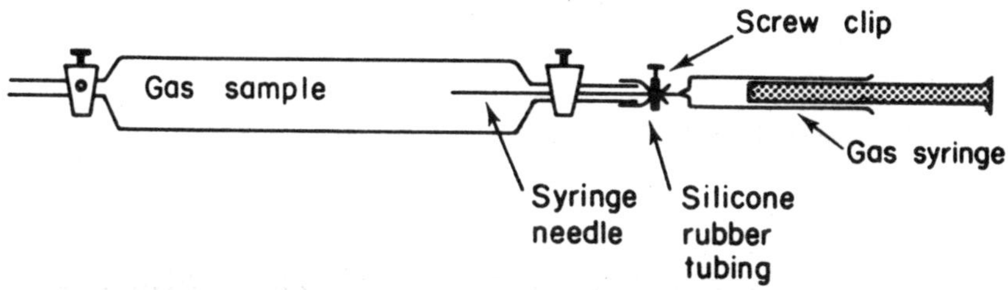

Fig. 4.4a. *Removing gas from a pipette*

The silicone rubber tubing, properly crimped by the screw clip, provides a gas-tight seal at the end of the cylinder. In order to maintain the integrity of the sample, the syringe should be flushed with the sample before the analysis sample is taken.

Π What do you consider to be the main problem, of taking a sample by the method described above, from a gas-sample that has been stored at atmospheric pressure? Can you suggest how this problem may be overcome.

The main problem associated with the taking of gas-samples from a fixed volume of gas under atmospheric pressure, is that if we assume that the cylinder holding the gas-sample is gas-tight, then we begin to create a vacuum inside the cylinder as we remove more and more sample. This means that every successive sample removed from the cylinder is at a slightly more negative pressure than the previous sample, and will achieve atmospheric pressure by sucking in air from the atmosphere as soon as the syringe is removed from the unit. The problem can best be overcome by pressurising the gas inside the cylinder before taking the sample. We can achieve this by attaching one end of the cylinder to a liquid displacement system as illustrated in Fig. 4.3a. The sample in the syringe will now have a positive pressure, and will rapidly attain atmospheric pressure as soon as the syringe is removed.

An alternative approach, is to remove the needle carefully from the cylinder and tap, and then close the tap. We can achieve a positive pressure for the gas inside the syringe by using the plunger. The sample of gas inside the syringe will attain atmospheric pressure on finally removing it from the silicone rubber attachment.

Constant-volume loops are often used for the sampling of gases, when glc is to be used as the analytical method. The loop is coupled to a multi-port valve, which is also attached to a cylinder of inert gas acting as the mobile phase in gas chromatography, and to the gas chromatographic column. The valve has two positions, a filling position and a sampling/analysis position. The gas flows through the valve are shown in Fig. 4.4b

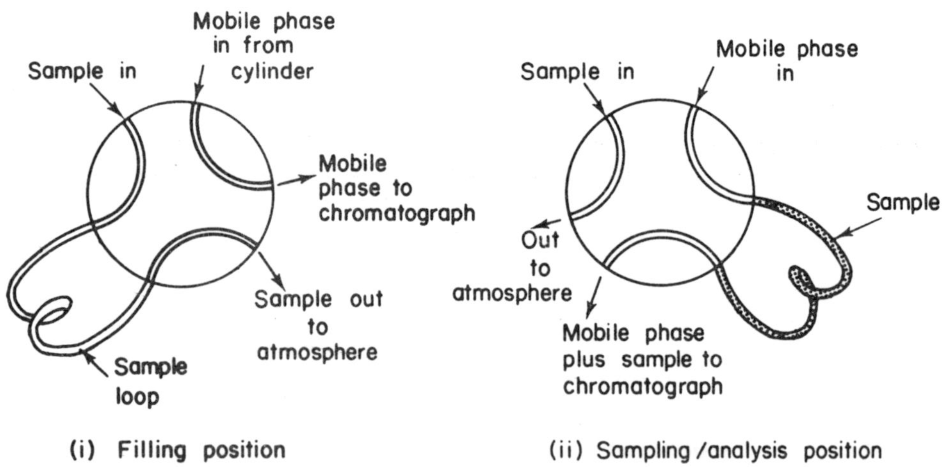

Fig. 4.4b. *Constant volume loop for gas sampling*

The gas to be sampled, is passed gently through the loop, both to flush out any gases previously held by the loop, and to fill the loop with the gas to be sampled. The valve is now rotated so that the gas sample contained within the loop (shown by the shaded area in Fig. 4.4b(ii)) is driven by the mobile phase, into the chromatography column, for chromatographic analysis.

If the gas sample was taken from a hot process-stream into a cylinder for transport to the laboratory, condensation of the less volatile components may have occurred between sampling and analysis. Before we take a sample for analysis therefore, the cylinder containing the sample should be warmed to the temperature at which the sample was originally taken, and allowed to remain at this temperature for several hours to allow the sample to reach equilibrium.

Objectives

When you have completed this part you will:

- know the main parameters to be considered when taking a sample,
- be able to choose appropriate sampling equipment for particular sampling situations,
- be aware of the extra problems caused by having to sample flowing systems,
- know how to prepare and take samples for analysis.

5. Standardisation and Calibration

There are very many analytical methods available to the scientist for obtaining qualitative and quantitative information about a specimen or a sample. Whilst a few of them appear to require no prior calibration or comparison with known standards, most of them do involve an initial calibration or standardisation.

It is the aim of this part of the Unit, to differentiate between those methods requiring calibration and those which do not, to consider the properties of standard substances, and to describe the best methods for calibration in particular situations.

5.1. GENERAL CONSIDERATIONS

Analytical methods for obtaining qualitative and quantitative results are often classified in a number of contrasting ways:

(*a*) chemical methods and physical methods,
(*b*) classical methods and instrumental methods,
(*c*) absolute methods and comparative methods.

None of these groupings, however, is perfect in classifying analytical methods. Physical and/or instrumental methods of analysis often involve much chemical pretreatment before the final physical or instrumental part of the procedure. Classical or chemical methods (eg gravimetry, titrimetry), although assumed to involve no instrumentation, do rely heavily on the use of the chemical balance, burettes, pipettes, etc, all of which are instruments of one sort or another.

For quantitative methods, the term *absolute* and *comparative* and are often used, and although not foolproof, these terms do provide the most satisfactory means of grouping quantitative analytical methods.

5.2. ABSOLUTE METHODS

Definition:

> '*These are methods involving a chemical reaction which achieve stoichiometric completion according to the chemical equation for that reaction.*'

As the above definition indicates, these are methods of analysis which involve carrying out a known chemical reaction, the reaction being expressed in terms of a chemical equation:

$$aA + bR \rightarrow A_aR_b \qquad (5.2a)$$

where A and R refer to the analyte and reagent respectively, and a and b refer to the number of moles of A and R respectively needed to produce a stoichiometric product. Eq. 5.2a, in strict terms, should be written as an equilibrium existing between the products on the right-hand side and reactants on the left-hand side. However as in all absolute methods we assume stoichiometric completion of the reaction, the equilibrium will lie completely over to the right-hand side of the equation as indicated by the arrow ($\rightarrow$).

There are very few analytical methods that can strictly be termed *absolute*.

Π Can you name some analytical methods which can be classed as absolute?

Fig. 5.2a below gives a complete list of those methods which can be classed as absolute. You may not have been able to name all of them, but I hope you managed to name gravimetry and titrimetry, which are the two best knowing examples.

Title of method	Property being measured
Gravimetry	Weight (mass)
Titrimetry	Volume of liquid
Gasometry	Volume of gas at STP
Coulometry	Charge transferred (Current × time)
Electrogravimetry	Weight (mass)
Thermogravimetry	Weight (mass)

Fig. 5.2a. *Absolute methods of analysis*

With the exception of thermogravimetry, the major advantages offered by all of these methods are those of *accuracy* and *precision*

Π Can you remember the difference between accuracy and precision?

Accuracy is defined as the closeness of the analytical result to the true figure, and should not be confused with *precision*, by which we mean the mutual closeness of a replicate set of results for the same analysis.

A set of replicate results may be very precise, but at the same time may be inaccurate.

Let us consider a simple example as illustration.

During the analysis of a sample of an alloy known to contain 19.80% of zinc, the following replicate set of results was obtained.

20.16%, 20.18%, 20.14% 20.15% and 20.19%

We should correctly consider the above set of analysis as fairly precise, indicating that the method used to obtain these results gave good reproducibility in the hands of the analyst who achieved them. However as the actual zinc content of the alloy was known to be 19.80%, we should at the same time, classify these results as somewhat inaccurate. The justification for our conclusion would be based upon a statistical examination of the results.

Under the best conditions, when using an absolute method it is possible to achieve both high accuracy and high precision. Indeed it is only with these methods that relative accuracy, and precision approaching 1 to 2 parts in 10^3 can be achieved. In terms of accuracy this is often referred to as ± 0.1 to 0.2% relative, meaning that the analytical result obtained is accurate to within ± 0.1 to 0.2% of the quantity of the analyte actually present.

You must bear in mind however, that in order to achieve these levels of accuracy and precision, a large quantity of sample must be available (ie not a few mg), and the proportion of analyte present in the sample must be substantial.

Π If an analyte is present at a level of 50.0% w/w, what would be an acceptable analytical result, given that it must be accurate to a least $\pm 0.2\%$ relative?

$$0.2\% \text{ of } 50 = \frac{0.2 \times 50}{100} = 0.1\%$$

$\therefore$ the result should lie between 49.9 and 50.1%

As indicated earlier, thermogravimetry cannot normally provide results with accuracy and precision equivalent to those of other gravimetric methods. This is because with this technique, in modern application one can analyse only very small quantities of a sample (less than 20 mg). The weight change occurring during the thermogravimetric procedure is measured on a calibrated potentiometric recorder, the accuracy of which in no way approaches $\pm 0.1\%$ relative.

Π When calcium carbonate is heated in air or nitrogen to about 700 °C, it loses carbon dioxide according to the equation:

$$CaCO_3 \rightarrow CaO + CO_2$$

By measuring the weight loss during heating, the percentage of calcium carbonate in the sample can be determined.

If a 10.0 mg sample containing $CaCO_3$, loses between 3.3 and 3.4 mg on being heated, calculate the percentage of $CaCO_3$ in the original sample. Relative molar masses of CO_2 and $CaCO_3$ are 44.0 and 100.0 respectively.

According to the equation for the reaction

$$1 \text{ mole of } CO_2 \equiv 1 \text{ mole of } CaCO_3$$

$$\therefore \quad 44.0 \text{ g } CO_2 \equiv 100.0 \text{ g } CaCO_3$$

For the results obtained:

$$3.3 \text{ mg } CO_2 \equiv 7.5 \text{ mg } CaCO_3$$
$$\text{and } 3.4 \text{ mg } CO_2 \equiv 7.7 \text{ mg } CaCO_3.$$

Remembering that the initial weight of sample taken was 10.0 mg, its percentage of $CaCO_3$ was between 75 and 77.

Π Calculate the relative accuracy for the analysis detailed in the above example.

The result was shown to be between 75 and 77% $CaCO_3$. The mean result is therefore 76% with a spread of ±1%. The relative accuracy is therefore $\frac{1\%}{76\%} \times 100 = 1.3\%$.

5.3. COMPARATIVE METHODS

Definition:

'Comparative analytical methods are those which require calibration against known standards, in order that accurate quantitative results may be obtained.'

Titrimetric and gravimetric procedures, as we have already seen, are generally considered to involve chemical reactions which achieve stoichiometric completion. If, however, a genuine equilibrium existed between the products and the reactants then in order to use the reaction for quantitative purposes, the position of this equilibrium would have to be determined. If this was so, the method would now become *comparative* and would involve an initial calibration. As it is, this situation rarely arises.

The majority of comparative methods are physical or instrumental techniques, whereby a single property of the analyte (solid or solution) is being measured. The property to be measured may:

— relate to the solution or solid as a whole, for instance, the measurement of viscosity or particle size;

— relate to a single component present in the solution or the solid, for instance the determination of the tin content of a mineral ore by X-ray fluorescence spectrometry;

— be inherent in the sample to be analysed, for instance, the determination of the ethanol content of a sample of blood by glc;

— be created *via* a suitable reaction, for instance, the colorimetric determination of the chromium content of a sample of steel, after oxidising the chromium to the dichromate ion ($Cr_2O_7^{2-}$).

SAQ 5.3a

Organo-sulphur compounds on combustion in oxygen, produce an equilibrium mixture of sulphur dioxide and sulphur trioxide according to the following equations:

$$\text{'S'} + O_2 \rightarrow SO_2$$

$$SO_2 + \tfrac{1}{2} O_2 \rightarrow SO_3$$

⟶

SAQ 5.3a (cont.)

The position of equilibrium depends upon experimental conditions, but a mixture is produced containing approximately 80% of the available sulphur as sulphur dioxide. This sulphur dioxide can be oxidised with iodine according to the following equations.

$$SO_2 + H_2O + I_2 \rightarrow SO_3 + 2H^+ + 2I^-$$

$$SO_3 + H_2O \rightarrow H_2SO_4$$

The reaction occurs rapidly and stoichiometrically.

Do you consider this analytical method for the determination of the sulphur to be absolute or comparative?

In all comparative methods there is a mathematical relationship that expresses the measured physical parameter as a function of analyte concentration,

$$Y = f(C) \tag{5.3a}$$

where Y is the measured parameter and C is the analyte concentration.

A calibration is thus required in order to establish in mathematical terms, the nature of the function relating Y to C.In the simplest case Y may be rectilinearly* related to C,

ie
$$Y = mC \tag{5.3b}$$

or in a more complex case Y may be related to log C

ie
$$Y = m \log C \tag{5.3c}$$

In Eq. 5.3b and 5.3c, m is termed the *proportionality constant* and has to be measured before the parameter can be converted into a concentration. The methods by which the value of m is found are termed *calibrations*, and these, together with some further examples of mathematical relationships are considered in greater detail later in this Unit.

* This is the strict usage: hereafter the simple terms 'linearly' or 'linear' will be used.

SAQ 5.3b

In a typical spectrophotometric determination, the relationship between measured absorbance and concentration of analyte being determined is given by:

Absorbance = k (concentration)

where k is the proportionality constant. ⟶

SAQ 5.3b (cont.)

If a solution containing 0.0100 mol dm^{-3} of the analyte gave an absorbance reading of 0.238, what would you expect the absorbance reading to be for a solution containing 0.0180 mol dm^{-3} of the analyte?

Also, if absorbance is dimensionless, what are the units in which *k* is measured?

For a calibration to be possible however, we must have available pure standards of known composition. These may be used directly, as for many organic compounds, or may be used to prepare solutions of known composition for metals.

For the analysis of trace quantities of metal ion present in aqueous solutions (trace meaning less than 100 ppm*), a standard solution of the metal ion containing 1000 ppm is often used. Standard solutions of these concentrations can be purchased from chemical suppliers or prepared in your own laboratory. They can be prepared either from the pure metal or from a pure compound of known composition.

(* ppm means parts per million and can be expressed in weight/volume terms as g cm^{-3}, mg dm^{-3} or g m^{-3}).

Π What weights of:

(*a*) pure copper metal,

(*b*) pure hydrated copper sulphate,

are required to prepare 1.000 dm^3 of a 1000 ppm solution of Cu^{2+}? The formula of hydrated copper sulphate is $CuSO_4.5H_2O$. Its relative molar mass is 249.5. The relative atomic mass of copper is 63.50.

(*a*) 1000 ppm $= 1000 \text{ mg dm}^{-3}$

$= 1.000 \text{ g dm}^{-3}$

Therefore, dissolve 1.000 g of the pure copper metal in dilute nitric acid and dilute the resultant solution to 1.000 dm^3.

(*b*) Hydrated copper sulphate contains 1 copper atom/mole.

Therefore to prepare 1.000 dm^3 of 1000 ppm Cu^{2+} solution we shall require:

$$\frac{249.5}{63.50} \times 1.000 \text{ g } CuSO_4.5\,H_2O = 3.929 \text{ g}$$

SAQ 5.3c

Calculate the weight of zinc required to prepare 250.0 cm^3 of a solution containing 1000 ppm of Zn^{2+}.

SAQ 5.3d

The amount of an analyte present in a material is known to be 25.00%. Within what limits would you expect to obtain a result for this analyte if the analysis is carried out;

(*i*) by using a gravimetric method,

(*ii*) by using a suitable optical method?

5.4 PRIMARY STANDARDS FOR USE IN CLASSICAL ANALYSIS

The term *Classical* methods of analysis is usually used for gravimetric and titrimetric techniques. As we have already seen these can be regarded as absolute methods of analysis.

In a typical gravimetric procedure, a pure crystalline material is precipitated from solution, filtered off, washed, and heated to constant weight. Provided the final precipitate is of precisely known chemical composition, then the final weight of the precipitate can be related to that of the analyte, *via* the chemical equation for the reaction. No initial calibration procedure is required, other than to satisfy oneself that the analytical balance is reading accurately.

In a titrimetric procedure however, the analyte in solution is treated with a titrant (solution of a reagent), according to a known chemical equation. The volume of titrant required to reach the equivalence point in the reaction is measured. Before the analyte concentration can be accurately determined, we need to know accurately, the concentration (molarity) of the titrant. This titrant will therefore have to be *standardised* against a pure substance of known composition, before it can be described as a *standard solution*. Substances used for this initial standardisation are called *primary standard substances* or *primary standards*.

5.4.1. Primary Standards

A perfect primary standard substance should satisfy the following requirements.

(*a*) It must be easy to obtain, to purify, to dry, and to preserve in a pure state.

Hydrated substances do not normally conform with these requirements owing to the difficulty of removing adsorbed water, without decomposition.

(*b*) It should not change in air during weighing.

It must not be hygroscopic or affected by the components of air.

(*c*) It should be capable of being tested for impurities which should normally not exceed 0.02% by weight.

(*d*) It should be of high relative molar mass in order to minimise weighing errors.

Modern analytical balances weigh accurately to ± 0.1 mg. Therefore, to achieve an accuracy of 0.1% relative, a sample weight greater than 0.1 g is required.

(*e*) It should be readily soluble (usually in water) under the conditions in which it is to be used.

(*f*) It should react instantaneously with the solute in the titrant solution, according to a single well-defined reaction.

(*g*) The titration error should be negligible, or easy to determine experimentally.

Assuming a coloured indicator is to be used to detect the end-point, then the titration error is usually the indicator error

In practice very few standard substances are perfect, and the analytical chemist is often forced to accept a compromise. A list of some suitable standards for titrimetry is given in Fig. 5.4a below.

Class of titration	Standard substances	
Acid-base	(di) sodium carbonate	Na_2CO_3
	disodium tetraborate decahydrate (Borax)	$Na_2B_4O_7.10\,H_2O$
	potassium hydrogen phthalate	$KH(C_8H_4O_4)$
	benzoic acid	$C_7H_6O_2$
Redox	(di) potassium dichromate	$K_2Cr_2O_7$
	potassium bromate	$KBrO_3$
	potassium iodate	KIO_3
	(di) sodium oxalate	$Na_2C_2O_4$
	arsenic(III) oxide	As_2O_3
Precipitation (silver halide)	silver	Ag
	silver nitrate	$AgNO_3$
	sodium chloride	NaCl
	potassium chloride	KCl
Complexometric (EDTA)	zinc	Zn
	magnesium	Mg
	copper	Cu
	EDTA (disodium salt)	$C_{10}H_{14}N_2O_8Na_2$

Fig. 5.4a. *List of some approved Primary Standards*

SAQ 5.4a

You are asked to prepare four litres of a standard solution of sulphuric acid (approximately 0.1 M) from an AnalaR sample of the concentrated acid. Which if any of the following procedures would be suitable?

[Concentrated sulphuric acid contains 96% w/v H_2SO_4; density conc. H_2SO_4 = 1.84 g cm^{-3}, $M_r(H_2SO_4)$ = 98.0]. ⟶

SAQ 5.4a (cont.)

(*i*) By using a measuring cylinder, transfer 22 to 23 cm^3 of the concentrated sulphuric acid into a large beaker containing about 2 litres of distilled water. Stir the solution continuously whilst adding the acid, and then allow the dilute acid to cool to room temperature before transferring it quantitatively to a 4 litre volumetric flask. Dilute the resultant solution to volume with distilled water.

(*ii*) By using a calibrated bulb pipette, transfer carefully 22.0 cm^3 of the concentrated sulphuric acid into a large beaker containing about 2 litres of distilled water. *Remainder of procedure as per* (*i*) above.

(*iii*) Weigh out accurately 40.333 g of concentrated sulphuric acid into a 4 litre beaker containing about 2 litres of distilled water. *Remainder of procedure as per* (*i*) above.

SAQ 5.4b

Which of the following reagents could be used to standardise an approximately 0.1 molar solution of dilute hydrochloric acid?

(*i*) Anhydrous disodium oxalate,

(*ii*) A solution of sodium hydroxide prepared from sodium hydroxide pellets,

(*iii*) Anhydrous disodium carbonate.

Objectives

Now that you have completed this part you will be able:

- to distinguish between analytical methods classed as absolute and comparative;
- to appreciate the advantages obtained by using absolute methods;
- to express comparative methods in terms of a simple mathematical relationship;
- to calculate the quantities of reagents required to prepare standard solutions containing ppm quantities;
- to list the main features of a primary standard and decide on the use of these standards.

6. Analytical Standards and Calibration Curves

There are many analytical methods that require some form of calibration before use in quantitative analysis. Although not the only way, a very important and popular method for this calibration is by means of a graph.

You will recollect from Section 5.3 that calibration procedures were introduced by stating that in all comparative methods, the measured parameter Y was a function of concentration. Let us now look in more detail at some ways in which this relationship exists and how it is used.

6.1. SIMPLE LINEAR FUNCTIONS

The general equation for a linear function, expressed in concentration terms is given by Eq. 6.1a.

$$Y = mC + b \tag{6.1a}$$

Where Y = measured parameter,
C = concentration of analyte,
m = proportionality constant,
b = a constant term (often considered as the experimental blank value).

A blank value can be defined as a measurement made on a sample, following an established procedure, but where the analyte was not added to the sample.

Fig. 6.1a shows Eq. 6.1a in graphical terms.

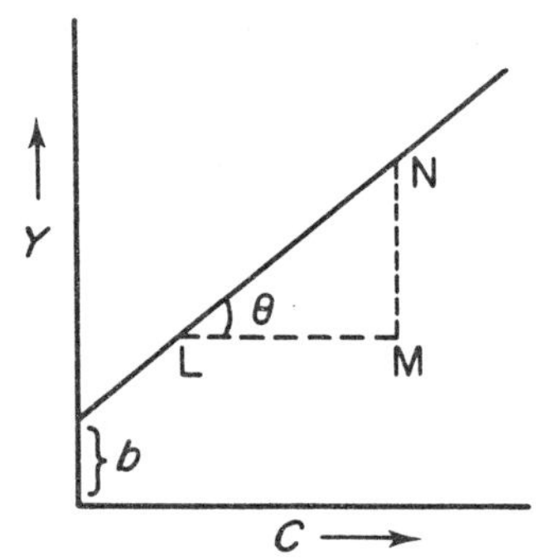

Fig. 6.1a. *Graphical representation of a simple linear function*

The proportionality constant m is given by the *slope* of the graph illustrated in Fig. 6.1a and can be calculated as below.

$$m = \tan\theta = MN/LM \qquad (6.1b)$$

When measuring LM and MN, always remember to relate these to the scales on your axes, and do not measure purely in terms of distance.

Π Fig. 6.1b shows a typical calibration plot for an analytical method obeying the equation $Y = mC + b$.

Calculate the slope of the line.

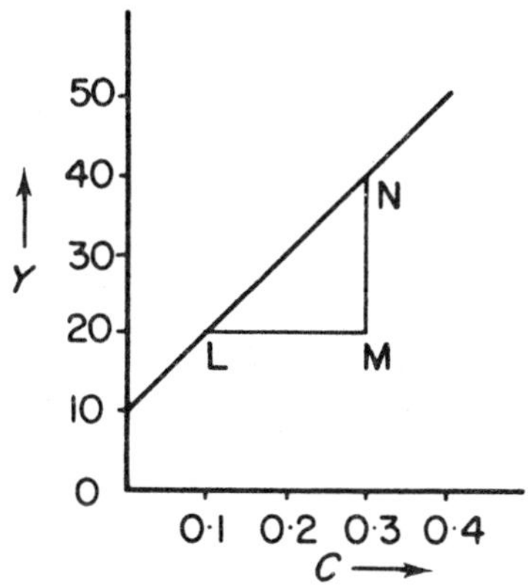

Fig. 6.1b. *Calibration plot for an analytical method*

From Fig. 6.1b as labelled

$$m = MN/LM = 20/0.20 = 1.0 \times 10^2$$

In the general form of the straight-line calibration plot as illustrated by Eq. 6.1a, the value of the intercept b may be *positive or negative or zero*. This intercept whether it be positive or negative can often be allowed for by subtracting the blank value from the standard and the sample value before respectively plotting or using the calibration graph. When, however, a positive or negative intercept remains after subtraction of the blank value, the analytical method is said to possess a *determinate error*.

Determinate errors are defined as *measurable errors which in theory have a definite value and which can be allowed for during the analysis*. They may be caused by the analyst, the instrument, or the method.

Π In a colorimetric determination of formaldehyde (methanal) the following results were used to construct a calibration graph:

Mass of formaldehyde (μg)	Absorbance
0	0.036
20.0	0.154
40.0	0.273
60.0	0.389
80.0	0.507
100.0	0.626

Is the method exhibiting a determinate error?

In order to see if there is a determinate error the blank value must be subtracted from all the sample values and then a graph of *corrected absorbance* against *mass of formaldehyde* should be constructed.

Mass of formaldehyde (μg)	Corrected absorbance
20.0	0.118
40.0	0.237
60.0	0.353
80.0	0.471
100.0	0.590

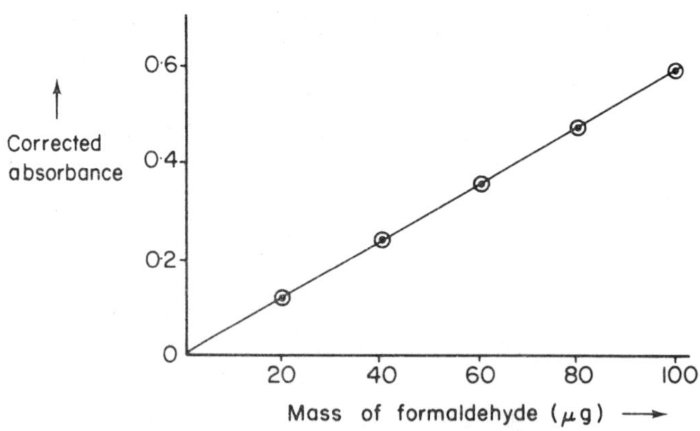

Fig. 6.1c. *Identification of determinate errors*

As all of the points lie on a straight line which can be extrapolated through the origin, the method has been shown, within experimental error, to have no determinate error. In fact the equation now becomes that originally shown in Eq. 5.3b,

ie $$Y = mC$$

It should be obvious that a positive intercept may arise from the presence of the analyte in the reagents used in the analytical method. It may be less obvious how a negative intercept can occur. One way, is in spectrophotometric measurement on solutions, when unmatched cells are used and the reference cell absorbs more radiation than the sample cell.

SAQ 6.1a

Fig. 6.1d below shows a section of a linear calibration plot. Is the sensitivity of the method, to which this calibration relates approximately:

(*i*) 2.10 μA ppm^{-1},

(*ii*) 2.81 μA ppm^{-1},

(*iii*) 0.54 μA ppm^{-1},

(*iv*) 0.48 μA ppm^{-1}? ⟶

SAQ 6.1a (cont.)

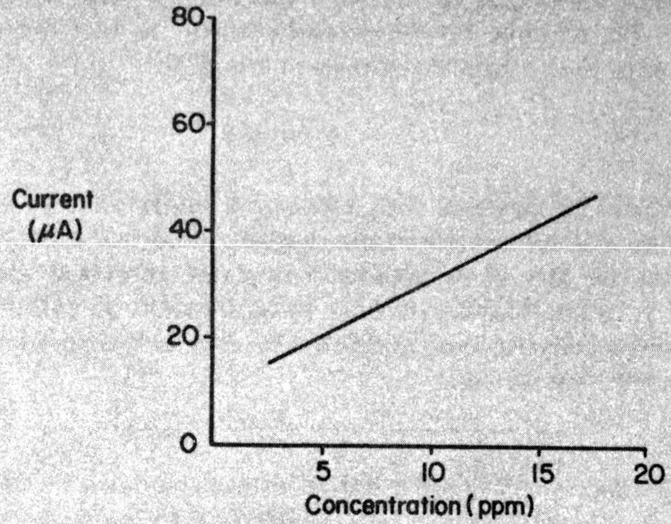

Fig. 6.1d. *Linear calibration plot for use in SAQ 6.1a*

We can easily demonstrate, in practical terms, the linearity as suggested by Eqs. 5.3b and 6.1a, by carrying out measurements on a range of standard solutions (or samples) all prepared from the pure analyte. We must be prepared however to accept that the results obtained may not necessarily all fall on a perfect straight line. This poor precision of results is normally due to an *indeterminate error* in the analytical method.

Definition of indeterminate error: '*Errors that arise from uncertainties in a measurement, they are unknown and cannot be controlled by the analyst. The effect will produce a scatter of results for replicate measurements which can be assesed only by statistical tests.*' (See the ACOL Unit: *Measurement Statistics and Computation* for a more comprehensive description and discussion). Given that our points do not all fall on a perfect straight line we need some means of finding the best straight line to draw through the experimental results obtained. This can be done by using *linear regression analysis*.

6.1.1. Linear Regression Analysis

The simplest procedure for linear regression analysis is that known as the *method of least squares*. It uses a mathematical formula to calculate the values of m and b in our linear equation, given that we have a set of data relating Y to C. The two values are calculated by using Eqs. 6.1c and 6.1d below:

$$m = \frac{\Sigma C \Sigma Y - n \Sigma CY}{(\Sigma C)^2 - n \Sigma C^2} \tag{6.1c}$$

$$b = \frac{\Sigma C \Sigma CY - \Sigma C^2 \Sigma Y}{(\Sigma C)^2 - n \Sigma C^2} \tag{6.1d}$$

where n = number of results available,

Σ = the sum of the designated functions of all of its n results.

When applying the method of least squares you will make a number of assumptions.

(*a*) That a linear relationship does indeed exist between the measured parameter Y and the analyte concentration C.

(*b*) That no significant error exists in the composition of the standard or in C. Thus the deviation of the points from an exact straight line are entirely due to the indeterminate error in the value of Y.

(*c*) That there are no pieces of data, which fall outside the normal statistical pattern, included in the calibration. These pieces of 'bad' data, as they are sometimes called, should be rejected, either after visual examination or after the use of statistical tests (see ACOL Unit: *Measurement, Statistics and Computation*).

The accuracy of a least-squares fit is a function of the number of data-points used in the calibration. A minimum of six points is recommended. However in this Unit, to reduce the tedium of the calculation, fewer results than the recommended number have been used.

Π The set of data listed in the table below relates to an analytical method obeying the following linear relationship.

$$Y = mC + b$$

Y	C
0.33	4.0
0.50	8.0
0.75	12.0
0.93	16.0

Before you commence this calculation, on the squared paper below, plot the four points given. Now attempt visually to construct the best straight line through these points.

By using the method of least squares, calculate the value of m and of b (constants), and hence superimpose a graph which represents the best straight line through the data-points given.

Compare the two straight lines.

Looking at Eq. 6.1c and 6.1d we see that the values we need to calculate initially are;

$$C^2, \quad CY,$$

and then

Y	C	C^2	CY
0.33	4	16	1.32
0.50	8	64	4.00
0.75	12	144	9.00
0.93	16	256	14.88
$\Sigma Y = 2.51$	$\Sigma C = 40$	$\Sigma C^2 = 480$	$\Sigma CY = 29.2$

$$\therefore \qquad (\Sigma C)^2 = 1600$$

Now $n = 4$

$$\therefore \qquad m = \frac{(40 \times 2.51) - (4 \times 29.2)}{1600 - (4 \times 480)}$$

$$= \frac{100.4 - 116.8}{1600 - 1920}$$

$$= \frac{-16.4}{-320}$$

$$= 0.051$$

and

$$b = \frac{(40 \times 29.2) - (480 \times 2.51)}{1600 - (4 \times 480)}$$

$$= \frac{1168 - 1204.8}{1600 - 1920}$$

$$= \frac{-36.8}{-320}$$

$$= 0.115$$

In order to construct the best straight line by using these values we return to the equation for that line, ie

$$Y = mC + b$$

we select the maximum value of C used in the calibration (ie C = 16), and calculate the value of Y which corresponds to this value

$$\therefore \quad Y = (0.051 \times 16) + 0.115$$

$$= 0.931$$

We now have two points through which we can construct our straight line.

$$Y = 0.115, \quad C = 0$$

$$Y = 0.93\ , \quad C = 16$$

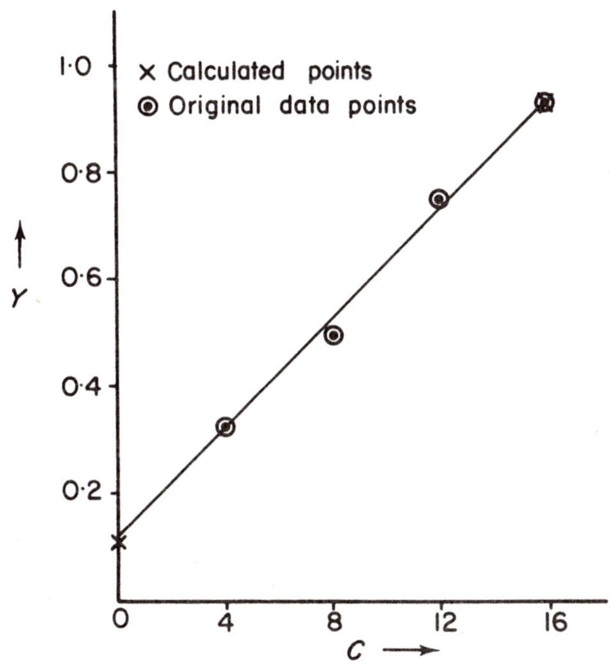

Fig. 6.1e. *Best straight line through data points*

As you can see from the example above, this calculation is somewhat tedious, but necessary where a rather imprecise set of results is obtained experimentally. Thankfully in the modern laboratory, this calculation is normally carried out on a computer. This reduces both the tedium, and the possibility of errors created by manipulating mathematical functions containing similar notations. Most computer programs which provide a least-squares fit also provide data for standard deviation, correlation coefficient, and possibly confidence limits. (See the Unit on Measurement, Statistics and Computation for the definitions and full description of these terms).

There is one final point to note before we leave this section – *never extrapolate a graph beyond the calibrated region.* As we shall see in the next section, there are many calibration plots that are linear only for low concentrations, and thus extrapolating a linear plot beyond the calibrated region may not be justified.

6.2. NON-LINEAR CALIBRATION PLOTS

Although most comparative methods obey a linear relationship as already described in Section 6.1, this obedience to linearity is often shown only over a relatively small concentration range. Go outside this range and negative deviation may occur as illustrated in Fig. 6.2a.

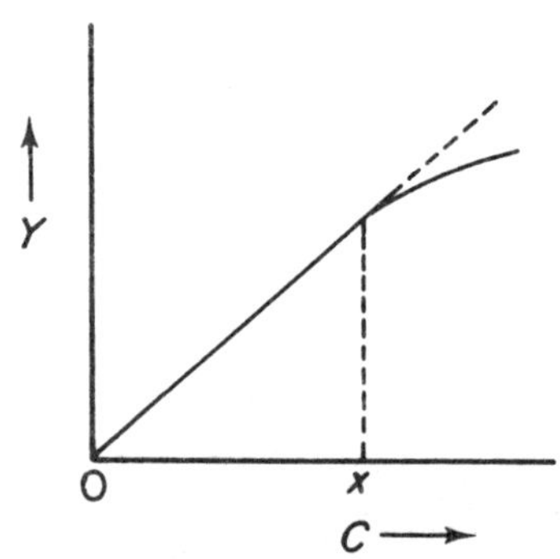

Fig. 6.2a. *Negative deviation*

From zero concentration to concentration (X) the linearity is obeyed, but above concentration (X) the system is showing negative deviation. Curvature does not affect the validity of a calibration plot, but does mean that more care must be taken in preparation, by using more standards to predict the curve accurately. The effect of curvature in calibration plots is often met in spectrophotometric methods involving emission or absorption of radiation. The curvature arises from instrumental features or from chemical changes that occur at higher concentrations and will be fully explained in a later Unit. The main problem with analytical methods which exhibit curved calibration plots is that the methods are not open to abridged calibration, standard addition, or internal standard procedures. All of these depend , as we shall see later in this Unit, on the existence of linearity between the measured parameter and concentration.

6.2.1. Transformation of Non-linear into Linear Functions

In analytical chemistry there are a number of useful techniques where the function relating Y to C is not initially linear, eg:

$$Y = k\,10^{-aC} \qquad (6.2a)$$

$$10^{aY} = kC \qquad (6.2b)$$

where Y = measured parameter,
C = concentration,
and a and k are constants.

Eq. 6.2a tells us that there is a reduction in the parameter Y with increase in concentration, with the opposite being true for Eq. 6.2b. The equations are shown graphically in Fig. 6.2b.

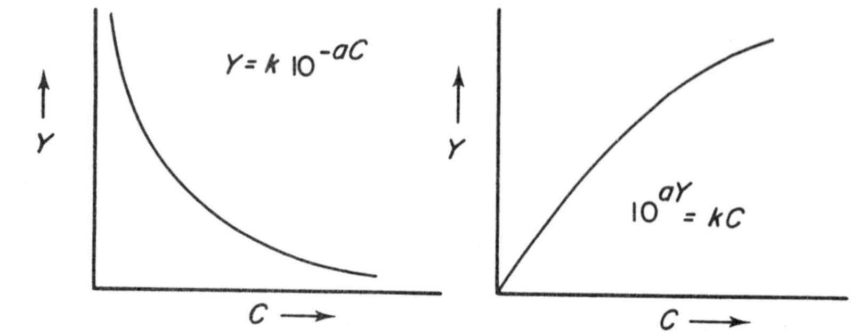

Fig. 6.2b. *Graphical representation of Eq. 6.2a and 6.2b*

We can, however, create a linear relationship from both of these equations by taking the logs of both sides.

Thus, taking logs for $Y = k\,10^{-aC}$ gives

$$\log Y = \log k - aC \qquad (6.2c)$$

and taking logs for $10^{aY} = kC$ gives

$$aY = \log k + \log C \qquad (6.2d)$$

The plots relating to Eq. 6.2c and 6.2d are illustrated in Fig. 6.2c below.

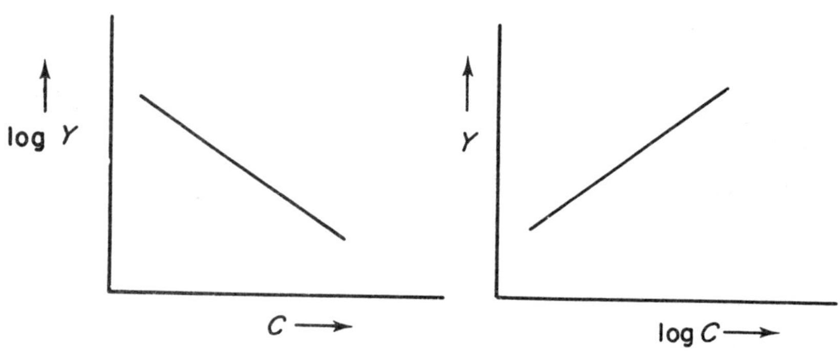

Fig. 6.2c. *Graphical representation of Eq. 6.2c and 6.2d*

The availability of modern scientific calculators has simplified considerably calibration involving logarithmic functions and their subsequent application.

Π The table below gives a set of calibration data for an analytical method obeying the relationship.

$$Y = k + m \log C$$

Concentration of analyte, C (ppm)	Measured parameter Y
100	0.150
200	0.173
300	0.186
400	0.195

If a sample solution gave a value of Y of 0.179, what was the concentration of the analyte in the sample?

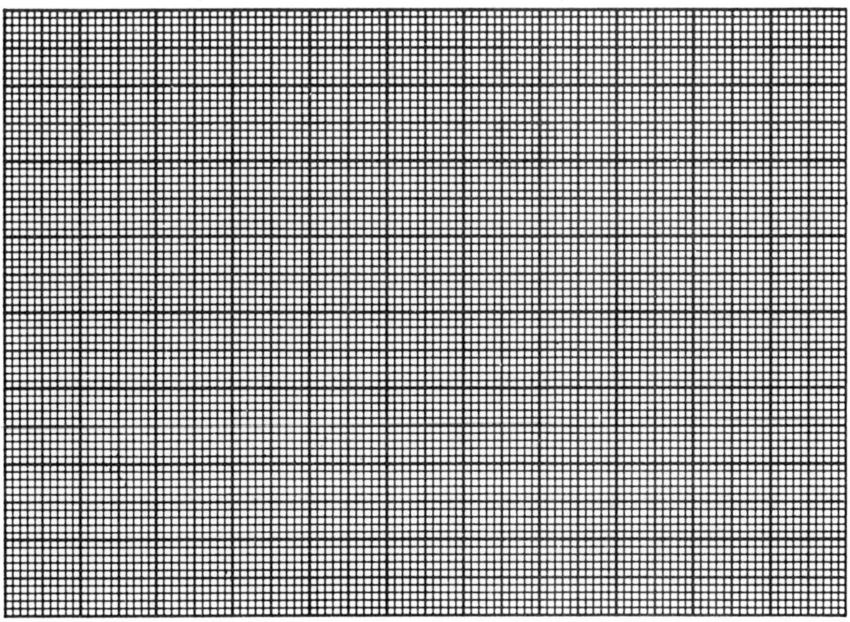

As we are told that the relationship is that Y is proportional to log C, we must first calculate log C.

C	log C	Y
100	2.00	0.150
200	2.30	0.173
300	2.48	0.186
400	2.60	0.195

We can now plot a Y vs log C graph

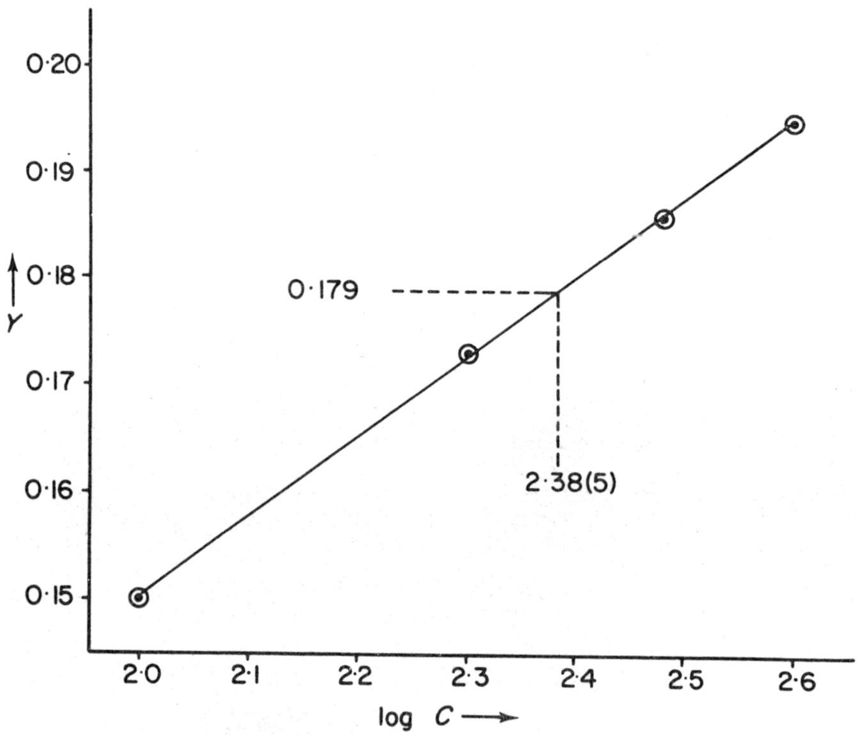

Fig. 6.2d. *Determination of analyte concentration (see text)*

From our graph we can now find the log C value corresponding to a Y value of 0.179. This is log C = 2.38(5)

By taking the antilog of this we can obtain C.

$\therefore$ Antilog of 2.38(5) = C = 243

The sample therefore contains 243 ppm of the analyte.

SAQ 6.2a

An analytical method is known to obey the relationship

$$Y = be^{-mC}$$

(*A*) Does this method exhibit a relationship whereby either

(*i*) *Y* increases with increasing concentration,

(*ii*) *Y* decreases with increasing concentration?

Attempt to answer part (*A*) of this SAQ before continuing to part (*B*).

(*B*) The following set of data was obtained from an analytical method thought to obey the experimental relationship quoted above.

⟶

SAQ 6.2a (cont.)

Y	C
2.26	2.00
1.97	4.00
1.67	6.00
1.45	8.00
1.23	10.00

What are your conclusions regarding this hypothesis?

Is the hypothesis

(*i*) definitely true,

(*ii*) probably true,

(*iii*) definitely false,

(*iv*) probably false?

6.3. ABRIDGED METHODS OF CALIBRATION

Where a linear calibration is either known or can be obtained as a result of linear transformation, it is sometimes possible to simplify the calibration and thereby to reduce the calibration time. This is achieved by reducing the number of standards required to construct the plot. There are three ways in which this can be done. In all these ways we must be confident of both our own ability to prepare accurate standards of known concentration, and that of the instrument to generate an accurate measurement.

6.3.1. Using a Single Standard

To use this procedure we must assume that the system *does not have a determinate error*, and that the reagents used give a *zero blank value*. The procedure can therefore be used only when the simplest equation ($Y = mC$) applies.

The concentration of standard chosen should relate to the maximum concentration where linearity still holds, or to the value of minimum error in the measured parameter; an absorbance between 0.4 to 0.5 when absorbance methods are being used.

There is in fact no need to construct a calibration plot, as a simple proportion relationship can be used to calculate the concentration of an unknown by reference to that of a known standard.

ie Concentration of unknown =

$$\frac{\text{Measured parameter of unknown}}{\text{Measured parameter of standard}} \times \text{Concentration of standard}$$

6.3.2. Using a Single Standard and a Blank

Where the procedure adopted *has no determinate error*, but does produce a *finite blank value*, then a blank determination must be carried out, and the blank value subtracted from all of the standard and sample measurements. The procedure for calibration will then be the same as that given in Section 6.3.1.

6.3.3. Using Two Standards and a Blank

When the system *possesses a determinate error* together with a *finite blank value*, three calibration measurements must be made: two standards and one blank. The standards chosen should be fairly widely spaced in terms of concentration. The high value should represent either the limit of linearity of the method or the limit of the expected analyte concentration. The low value should relate approximately to the minimum concentration commensurate with the accuracy of the measurement. Both values should be plotted on a graph after the blank value has been subtracted, as illustrated in Fig. 6.3a.

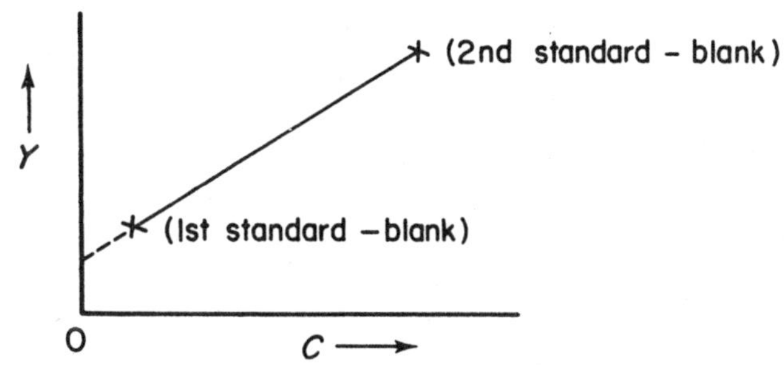

Fig. 6.3a. *Calibration using two standards and a blank*

Π In a spectrophotometric procedure, the following calibration results were obtained.

Analyte concentration (ppm)	Absorbance (A)
150	0.246
300	0.417
Blank	0.020

By assuming that a linear relationship exists between absorbance and the analyte concentration, calculate the concentration of the analyte in a sample, given that the measured absorbance was 0.317

Before plotting a graph of absorbance *vs* concentration we must first subtract the blank absorbance value of 0.020 from the two standard absorbances and the sample absorbance.

Analyte concentration (ppm)	Corrected absorbance (A')
150	0.226
300	0.397
?	0.297

We can now construct our calibration plot with the two standard points, and hence obtain the concentration of analyte present in the unknown.

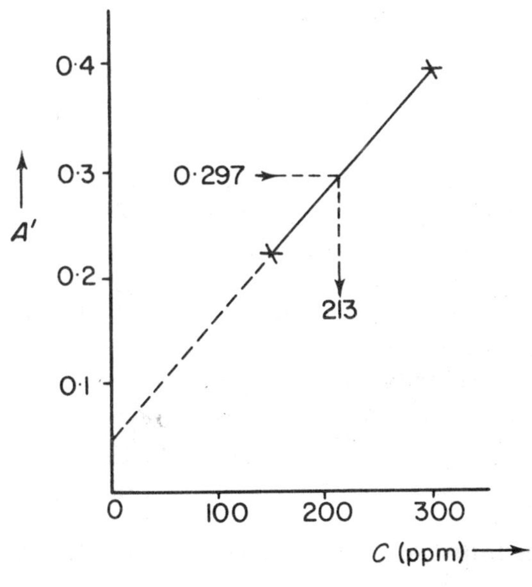

Fig. 6.3b. *Calibration graph*

From the graph, the concentration of analyte in the unknown is *213 ppm.*

6.4. CHOICE OF STANDARD MATERIALS

In Section 5.4.1 we considered in detail the properties required of a material in order that it could be called a primary standard. Unfortunately, there are very few compounds that conform precisely to all these requirements, and therefore it is necessary to have a more lenient attitude to compounds that can serve as standards. Some, but by no means all of these substances we tend to use, are termed *secondary standards*.

A secondary standard is, by definition, a substance which can be dissolved (usually in water) to form a stable solution, and which can be subsequently standardised against a primary standard. These secondary standards are often used in titrimetric analysis (eg sulphuric acid is a secondary standard) or for the preparation of standards to be used with comparative techniques. With comparative techniques, however, we can go one step further, and use materials and compounds as standards, without the necessity of an initial standardisation.

Bearing in mind that comparative techniques are normally accurate only to $\pm 1\%$ relative, substances can be used as standards that are guaranteed to at least 99.0% purity. Therefore for comparative procedures, we can relax the stringent requirements we make of a material when we intend to use it for standardisation.

Thus a much wider range of compounds, both inorganic and organic, in solid, liquid, or gaseous form, now become possible standards. The vast majority of calibrations in comparative analytical techniques are carried out by using a source of the analyte itself, provided it has the following properties as a standard substance.

Properties of a Substance for Use in Comparative Techniques

- It must be available in a known state of purity
- It must be stable during weighing.
- It should be readily soluble (when required) under the conditions in which it is to be used.

SAQ 6.4a

Which of the following properties are important in choosing a suitable standard for calibration in a comparative analytical technique?

(*i*) The standard substance should not react with any components normally present in air.

(*ii*) Impurities present in the standard substance should not normally exceed 0.1%.

(*iii*) The standard substance should be of a high relative molar mass in order to minimise weighing errors.

(*iv*) The standard substance should be readily soluble in the solvent under the conditions in which it is to be used.

(*v*) The standard substance does not necessarily have to be of an accurately known formula, so long as the percentage of the analyte constituent is accurately known.

(*vi*) The standard substance should be present in an anhydrous state.

6.4.1. Optimum Concentration Range

We have already seen in Section 6.2 that calibration plots do not of necessity have to be linear, although it is to our advantage if they are.

Π What is the main advantage offered by linear calibration plots?

The main advantage of linear calibration plots is that fewer standards are required to produce the calibration. In some instances, as we saw in Section 6.3.1, only a single standard is required. With curved calibrations plots, however, it is always necessary to use a number of standards so as to determine the calibration graph accurately.

One problem which is inherent when using curved calibrations is that the curvature of the line may increase with concentration. This is illustrated in Fig. 6.4a.

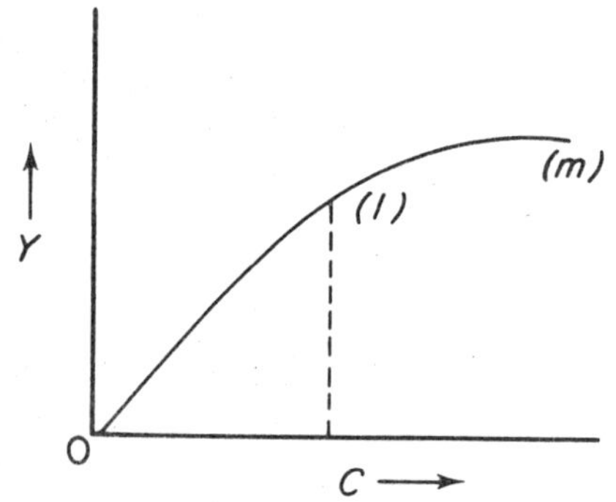

Fig. 6.4a. *Increase of curvature with concentration*

We can divide the curve shown in Fig. 6.4a into two parts: that between 0 and concentration l which is essentially linear, and that between concentration (l) and (m) which becomes progressively more non-linear with increase in concentration. In choosing an optimum calibration range we must take into account two important factors:

- the overall accuracy with which the parameter Y may be measured – a factor inherent in the techniques being used;
- the slope of the curve (dY/dC) at any particular point on the plot.

Every comparative method of analysis, particularly those which are instrumentally based, exhibits a range of concentrations over which the greatest accuracy of measurement can be obtained. It is not part of this Unit to discuss the reasons for this phenomenon, but it will serve as an illustration to select one technique where this effect is well recognised.

Example. Consider the technique of atomic absorption spectroscopy (Aas)

Aas is an important spectroscopic technique for the determination of trace metals in solution. The technique measures absorbance of atoms and it can be shown theoretically that the maximum accuracy for measuring an absorbance is when the absorbance value is about 0.4 with acceptable practical accuracy being achieved between absorbance values of 0.2 and 0.9.

Aas is also a technique prone to curved calibration plots as illustrated by the calibration for lead in Fig. 6.4b and shown in Fig. 6.4c below.

Concentration (C) of lead (ppm)	Absorbance (A) at 217.0 nm
5.00	0.142
10.0	0.249
15.0	0.337
20.0	0.427
25.0	0.507
50.0	0.762
100.0	0.994
150.0	1.089

Fig. 6.4b

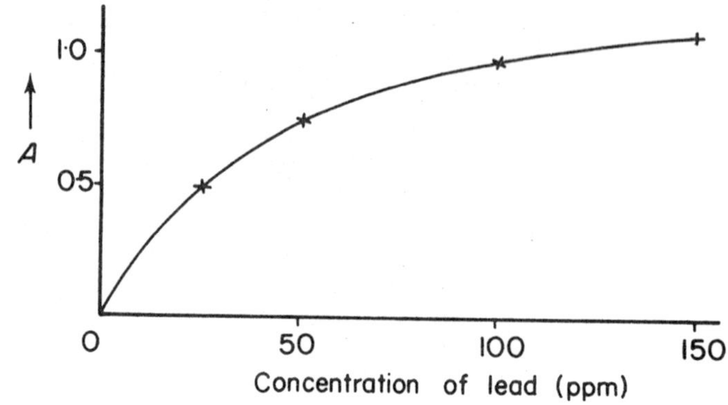

Fig. 6.4c

Π By referring to the data in Fig. 6.4b and shown graphically in Fig. 6.4c state what you consider to be a maximum acceptable concentration range for the analysis of lead under the conditions for which these calibration data were obtained.

As we have not been given the accuracy within which each absorbance value can be measured, the answer to this question is partly subjective. However we can first obtain the slope of the graph between each of the pairs of data points.

C (ppm)	A	$\Delta(A)/\Delta(C)$
5.00	0.142	
		0.021
10.0	0.249	
		0.018
15.0	0.337	
		0.018
20.0	0.427	
		0.016
25.0	0.507	
		0.010
50.0	0.762	
		0.005
100.0	0.994	
		0.002
150.0	1.089	

Strictly we should have obtained the precise slope of the curve at each data point. This however is a time-consuming process. The method adopted of converting the curve into a series of straight lines will serve to illustrate the effect of changing slope.

The slope obtained is an approximate measure of the sensitivity of the method between successive pairs of data points. The results show that the sensitivity has decreased by approximately 25% up to 25 ppm, by about 50% up to 50 ppm and by about 75% up to 100 ppm. Also, given the fact that the most accurate measure for absorbance is about 0.4, with acceptable accuracy up to absorbances of about 0.9, the method should probably not be used above concentration levels of 50 ppm. The optimum calibration range for this method is probably 0 to 25 ppm, with a possible extension to 50 ppm, especially if more data points between 25 and 50 ppm are included.

There are some analytical techniques for which linear calibrations extend over a very wide concentration range without significant loss of accuracy and precision. The optimum choice for calibration purposes should then relate to the expected concentration range of the analyte.

6.5. PREPARATION OF CALIBRATION GRAPHS

There are two main methods for the preparation of calibration graphs:

— preparation of separate standards,

— '*in-situ*' calibration

6.5.1. Method of Separate Standards

Of the two methods, this one has universal applicability and can be applied equally to analytical methods involving measurement of an analyte in a solution or in a solid matrix. Remember that there are a few techniques where analytical measurements are made directly on a solid matrix, eg X-ray fluorescence, arc emission spectroscopy.

The method involves the preparation of a set of solutions or homogeneous powders containing increasing amounts of the standard analyte. For liquid samples, we would normally dilute all the prepared standard solutions to the same known volume. However this is not strictly necessary, provided we know the concentration of the standard in each of the prepared solutions.

For solid matrices (powders mainly) we would find difficulty in preparing standards by this method. Standards are therefore prepared containing known percentages (w/w) of the analyte in a suitable matrix.

Π You are asked to prepare a set of a aqueous ethanol standards (100.0 cm^3 of each) containing up to 5% v/v of ethanol, from a sample of pure ethanol provided. Suggest a suitable range of concentrations for these standards and a procedure for their preparation. Remember, that ethanol is more volatile than water.

When preparing standard aqueous solutions of volatile reagents we must take the precaution of adding the reagent to a flask already containing water. In this way we reduce considerably the loss of the reagent as vapour, when the flask containing the reagent is filled to volume.

Given that we are asked to prepare a set of standard solutions up to 5% v/v of ethanol, this probably means 4 or 5 solutions spread over the concentration range 0 to 5%. We should avoid fractional percentage quantities (ie 1.25, 2.50, 3.75%) as these tend to reduce the accuracy with which the solutions are prepared. A suitable choice is therefore a set of standards containing 1, 2, 3, 4 and 5% v/v of ethanol in water, together with a water blank.

These are best prepared by adding 1.00, 2.00, 3.00, 4.00 and 5.00 cm^3 respectively of pure ethanol to 100.0 cm^3 graduated flasks containing approximately 50 cm^3 of water. Each resultant solution is then diluted to the required volume and shaken to mix the components, the best method for delivery of the ethanol would be by calibrated bulb pipette. For the most accurate work the pipettes used should be *calibrated with ethanol* at a suitable temperature, eg 20 °C.

6.5.2. Method of '*in-situ*' Calibration

There are a few analytical methods where a calibration can be obtained by using a single solution, to which we can add sequentially, increasing amounts of the standard analyte. This is known as '*in-situ*' calibration. In order to produce a calibration by this method we should take initially a known volume of the blank solution or reagent, and place it in the analytical cell or instrument. Having taken a blank or background measurement on this solution, we can now add to it known volumes of standard analyte solution, so as gradually to increase the concentration of the analyte being determined. A measurement is taken after each addition. Although not absolutely necessary, it is preferable if the volumes of standard solution added are very small so as not significantly to change the total volume of solution in the analytical cell or instrument

Π In the polarographic determination of Cd^{2+} a calibration graph was constructed by using the following procedure.

10.0 cm^3 of dilute acid were pipetted initially into the polarograpic cell. To this situation, were added 10.0 μl aliquots of a standard solution of cadmium, containing 1000 ppm of Cd^{2+}. Assuming that the dilute acid contained no cadmium ions, what were the concentrations in ppm of Cd^{2+} in the polarographic cell after 10, 20, 30, 40 and 50 μl respectively of the standard cadmium solution had been added?.

$[1\ \mu l \equiv 10^{-6}\ l \equiv 10^{-3}\ cm^3 \equiv 1\ mm^3]$

Given that 1 ppm can be expressed as 1 $\mu g\ cm^{-3}$, the most appropriate volume units to apply are cm^3.

Now 10 μl = 0.010 cm^3 and contains 0.010 × 1000 μg of Cd^{2+}

$$= 10.0\ \mu g$$

The solution containing this weight of Cd^{2+} was diluted to 10.0 cm^3 with the acid originally in the polarographic cell, producing a solution of concentration 1.0 ppm.

By similar calculations the remaining standard concentrations are shown to be 2.0, 3.0, 4.0 and 5.0 ppm respectively.

It is important to note that the method of '*in-situ*' calibration can be used only when the analytical method:

(*a*) does not cause a significant reduction in the concentration of the analyte during the measuring process;

(*b*) does not lead to a loss in volume of the solvent present.

SAQ 6.5a

A colorimetric method is to be used for the determination of trace quantities of palladium present in aqueous solutions. You have followed the recommended procedure, which involved the preparation of separate standard solutions and have thereby obtained the calibration data below.

Mass of palladium (W) in 50.0 cm^3 of measured solution (μg)	Absorbance (A)
5.0	0.225
10.0	0.315
15.0	0.375
20.0	0.470
25.0	0.550
Blank (0)	0.075

Attempt to answer the following questions related to the analytical method described above.

(*i*) Assuming that a linear relationship exists between absorbance and weight of palladium in solution, construct the best straight line through the data-points given. What is the exact equation for the relationship?

→

SAQ 6.5a (cont.)

(*ii*) Is the analytical method exhibiting:
- a determinate error;
- an indeterminate error;
- both a determinate and an indeterminate error?

(*iii*) A sample containing an unknown quantity of palladium is analysed by the analytical method, giving the following results.

Sample absorbance = 0.615

Blank absorbance = 0.070

What is the weight of palladium in μg in the sample analysed? Comment upon the validity of the results obtained.

(*iv*) Do you consider that the calibration could alternatively have been obtained by using an '*in-situ*' method?

(*v*) How many of the three abridged methods of calibration could be used with this analytical procedure? Suggest which they are and give reasons for your choice.

SAQ 6.5a

6.6. MATRIX MATCHING

When developing or choosing an analytical method we should always emphasise strongly the *selectivity or specificity* achievable with that method. Selectivity or specificity refers to the ability of the method to determine the analyte without interference from the other substances present in the analyte matrix. Although it is relatively rare to find a truly specific method for a single analyte, we can avoid many interferences by the correct choice of standards and conditions for a particular analysis. This process is often termed *Matrix Matching of Standards.*

In matrix matching we prepare the standards so as to match, as closely as possible, the sample matrix in which the analyte is to be determined. This in simple terms, may mean the use of the same anionic matrix as in the sample, in the determination of a metal ion, or in a more complex case the exact matching of the other substances known to be present in the sample. When the matrix is highly complex, the method of standard addition would probably be preferred (see Part 7 of this Unit).

6.7. BRACKETING OF STANDARDS

In our consideration of calibration methods, we have assumed that the analytical method obeys a regular relationship (ie a straight line or a smooth curve). Although we have emphasised the necessity of increasing the number of calibration standards when a curved calibration exists, for economic reasons if no other, we are unlikely to use more than 5 or 6 standards in order to plot the calibration graph. The best curve is then constructed through these points by using a 'flexi-rule' or other suitable curve drawing device. Inevitably this can lead to a slight loss in accuracy when the calibration is finally used for analysis.

The alternative procedure therefore, which is sometimes favoured, involves the *bracketing of standards* around the analysis value – one standard being slightly higher than the analytical value and the other slightly lower. The procedure which would be adopted is summarised as follows.

(*a*) Analyse the sample by a given procedure for which a calibration graph had previously been prepared. Calculate the analyte concentration.

(*b*) Prepare two standards, one slightly less and the other slightly greater than the analyte concentration and analyse these by using the same procedure.

Provided the two standards differ only slightly in concentration it is feasible to assume that over this narrow concentration range a linear relationship exists between the measured parameter and analyte concentration. The calculation is performed as illustrated in the example given below.

Example

Consider two standard solutions containing 15.0 and 16.0 ppm respectively of the analyte. After analysis by a given procedure the absorbance of the two solutions was found to be 0.300 and 0.310 respectively. If an unknown sample containing the same analyte gave an absorbance of 0.304, what is the concentration of the analyte in the unknown?

As we are assuming that a linear relationship exists between the two standards, we can average the two values giving an analyte concentration of 15.5 ppm $\equiv$ an absorbance of 0.305.

$\therefore$ absorbance of 0.304 = 0.304/0.305 × 15.5 ppm = 15.4 ppm.

Advantage

Given that the analytical method being used is very precise, then the accuracy of the determination will be increased.

Disadvantage

We do need to carry out more measurements than would be normal with a standard calibration technique. The initial figure obtained from the calculation curve has to be fairly accurate in order that we can chose correctly the concentrations of the two standards to be bracketed around that of the sample.

List of Objectives

Now that you have completed this part you will be able to:

- define any straight-line calibration graph by a simple mathematical relationship, and construct the best straight line through a given set of data points;

- use non-linear calibration graphs, and be able to convert some non-linear functions into linear functions;

- appreciate the value of abridged methods of calibration when there is linear relationship.

- choose a suitable calibration standard for use with a comparative method, and to select an optimum range for the calibration;

- appreciate the effects of determinate and indeterminate errors.

- distinguish between the seperate-solution and the *in-situ* method for preparing calibration graphs.

7. Standard-addition Methods

In Section (6.6) of this Unit we considered the importance of attempting to match the composition of the calibration matrix to that of the sample. This is necessary, as there are many instances when the signal given by the analyte can be either enhanced or decreased by the presence of other components of the matrix. Unfortunately, it is often not feasible to take into account all the components likely to be present in the sample matrix. One method of attempting to overcome this problem is to calibrate and analyse within the sample matrix itself. This procedure is known as the *method of standard addition.*

The complexity of the standard-addition method may vary from one analytical technique to another. However, the procedure adopted always involves a set pattern of events.

(*a*) Preparation of the sample to be analysed by the addition of buffers, reagents, etc, and then measuring the physical parameter used for the analysis, (eg absorbance, current, potential),

(*b*) Addition to this sample of a known quantity of a standard substance containing the analyte, mixing the two together and remeasuring the physical parameter.

The increase in the value of the physical parameter is related to the quantity of the standard added, and therefore effectively acts as a means of *in-situ* calibration.

Before we consider in detail how the technique is used, three important points must be stressed

- *The method of standard-addition can be used only when a straight-line (linear) relationship exists between the physical parameter Y and the concentration of the analyte C.*

- When the analytical method has a determinate error of constant value, ie a constant term in the linear equation, the value of the slope of the straight line, m, must be known. Thus, when the analytical method involves relationships of the type:

  $$Y = mC + b$$

  or $$Y = m \log C + b$$

  the value of m must be known.

- The sample solution containing the analyte must not contain any interfering species that can masquerade as the analyte. Species of this type may or may not give rise to a linear relationship between Y and C_{int}. If they do, however, it is most unlikely that the proportionality constant m will be the same as that between Y and C_{anal}.

7.1. METHOD OF SINGLE STANDARD-ADDITION

The method involves adding a single standard of known concentration to the analyte solution under investigation and measuring the increase in the physical parameter caused by the addition of the standard. The analyte concentration is calculated by the use of simultaneous equations as illustrated below.

Let us suppose that in the method being used Y is related to C by the simplest of linear equations

$$Y = mC$$

Two measurements of Y are made:

(*a*) Y_o is that made on the sample solution *before* the addition of the standard;

(*b*) Y_i is that made on the resultant solution *after* the addition of the standard.

$$\therefore \quad Y_o = mC \tag{7.1}$$

and

$$Y_i = m(C + C_s) \tag{7.2}$$

where C_s is the increase in concentration of the analyte caused by the addition of the standard.

As most analyses involving standard addition relate to solutions, C_s must be measured either in molar or more likely in wt/vol terms. You must remember, however, that C_s depends on the total volume of solution undergoing analysis after the addition of the standard. Therefore, if for instance a volume of 1.0 cm^3 containing the standard at an initial concentration of 1.0 mg cm^{-3} is added to 25.0 cm^3 of a sample solution, the value of C_s would be:

$$\frac{1.0}{26.0} = 0.038 \text{ mg cm}^{-3}$$

Finally you should also note that the units for C will be dictated by your choice of units for C_s.

We now have two Eq. (7.1) and (7.2) and two unknowns, m and C. We can therefore solve these equations as follows.

From Eq. (7.1)

$$m = \frac{Y_o}{C}$$

We can substitute this value for m in Eq. (7.2).

$$\therefore \quad Y_i = \frac{Y_o}{C}(C + C_s) \tag{7.3}$$

$$= Y_o + \frac{Y_o C_s}{C} \qquad (7.4)$$

Rearranging Eq. (7.4) we get

$$Y_i - Y_o = \frac{Y_o C_s}{C}$$

$$\therefore \qquad C = \frac{Y_o C_s}{Y_i - Y_o} \qquad (7.5)$$

$(Y_i - Y_o)$ represents the increase in the physical parameter, and Eq. (7.5) is the general form of the equation for standard addition. However, in this equation we assume that *when the standard was added to the sample solution no significant dilution occurred*.

Dilution effects should be assumed significant when the volume of sample solution, as a result of standard addition, increases by more that 1%.

Π In the determination of caffeine by hplc a standard addition method was adopted.

A 5.00 cm^3 aliquot of a sample containing an unknown quantity of caffeine was added to each of two 25.0 cm^3 volumetric flasks. The contents of the first of these flasks were diluted to volume with water. To the second flask was added 5.00 cm^3 of a standard caffeine solution containing 0.500% w/v of pure caffeine. The resultant solution was also diluted to volume with water. The two solutions were then analysed by hplc, the peak areas in each chromatogram due to caffeine, being measured by using an electronic integrator. The results obtained were as follows.

	Peak area (mm^2)
Sample	1049
Sample + Standard	1965

By assuming that there is a linear relationship between peak area and concentration within the concentration range examined, calculate the percentage of caffeine in the unknown sample.

As both solutions analysed by hplc were diluted to the same volume, Eq. (7.5) can be used for this calculation.

5.00 cm^3 of the 0.500% w/v standard solution of caffeine was diluted to 25.0 cm^3

$$\therefore \quad C_S = 0.500 \times \frac{5.00}{25.0} \% \text{ w/v}$$

$$= 0.100\% \text{ w/v}$$

Now

$$C = \frac{Y_O C_S}{Y_i - Y_O} \% \text{ w/v}$$

$$= \frac{1049 \times 0.100\% \text{ w/v}}{1965 - 1049\% \text{ w/v}}$$

$$= \frac{104.9}{916}$$

$$= 0.115\% \text{ w/v}$$

Now as C_S was related to the concentration of the standard in 25.0 cm^3 of the analysed solution, so C must be related in the same way. But the sample containing the unknown quantity of caffeine was initially diluted 5.00 cm^3 $\rightarrow$ 25.0 cm^3 before analysis. Therefore the value of C must be multiplied by the dilution factor in order that the answer may relate to the original sample.

$\therefore$ Concentration in the original sample

$$= 0.115 \times \frac{25.0}{5.00}$$

$$= 0.575\% \text{ w/v}$$

SAQ 7.1a

A single standard addition procedure is to be used in an analysis involving a logarithmic relationship. If the two equations, before and after standard addition, are

before addition, $Y_o = b - m \log C$

after addition, $Y_i = b - m \log (C + C_s)$

(*i*) develop an equation which relates C to the other parameters in the equations,

(*ii*) having developed your equation, what extra information do you require in order to calculate C?

You may assume there are no significant dilution effects on addition of the standard.

We must now consider the effect of addition of the standard significantly altering the volume of the solution undergoing analysis.

We can now no longer consider the value of C to remain constant for the two simultaneous equations. Because the volume of solution containing the analyte has increased after addition of the standard, the value of C will have decreased proportionally.

After addition of the standard therefore, the value of C must be multiplied by a factor that takes into account the relative volume before and after addition of the standard.

Thus C now becomes

$$C \times \frac{V}{V + v}$$

where V is the original volume of solution containing the analyte, v is the increase in volume caused by addition of the standard.

Eq. (7.2) therefore becomes

$$Y_i = m[CV/(V + v) + C_s] \tag{7.6}$$

We now solve the simultaneous Eqs. (7.1) and (7.6) to obtain an equation relating C to the other parameters in the equation.

As m represents an unknown quantity, it is best to eliminate this initially from the two equations.

Eq. (7.1) is

$$Y_O = mC$$

or

$$m = Y_O/C$$

Substituting for m in Eq. (7.6) gives

$$Y_i = \frac{Y_O}{C}\left[C\left(\frac{V}{V + v}\right) + C_s\right]$$

$$= \frac{Y_O V}{(V + v)} + \frac{Y_O C_s}{C} \tag{7.7}$$

Rearranging Eq. (7.7) gives

$$Y_i - \frac{Y_O V}{(V + v)} = \frac{Y_O C_s}{C}$$

and therefore C is given by Eq. (7.8)

$$C = \frac{Y_O C_s}{\left[Y_i - \frac{Y_O V}{(V + v)}\right]} \tag{7.8}$$

By multiplying both the numerator and denominator of the right-hand side of Eq. (7.8) by $(V + v)$ we get finally Eq. (7.9).

$$C = \frac{Y_o C_s (V + v)}{Y_i (V + v) - Y_o V} \tag{7.9}$$

I think it a good idea at this stage briefly to restate the two equations which relate to the method of single standard-addition.

$$C = \frac{Y_O C_s}{Y_i - Y_o} \tag{7.5}$$

for use where no significant dilution occurs, and

$$C = \frac{Y_o C_s (V + v)}{Y_i (V + v) - Y_o V} \tag{7.9}$$

for use where significant dilution does occur.

Because of the similarity of the parameters within these two equations, I should not advise you to try to remember them. It is better to derive each equation from first principles as detailed in the text, whenever they are required.

SAQ 7.1b

In the technique of anodic stripping voltammetry, the peak height of the anodic wave is proportional to the concentration of analyte according to the relationship:

$$i_p \propto C$$

where i_p refers to peak height and providing that all experimental conditions are held constant.

The following data relate to the determination of traces of lead in drinking water.

Volume of water sample analysed	= 10.0 cm^3
Initial peak height of the lead wave	= 1.56 μA (i_{po})
Volume of standard added	= 1.00 cm^3
Concentration of lead in the standard	= 0.100 mg dm^{-3}
Peak height of the lead wave after addition of the standard	= 3.79 μA (i_{pi})

(*i*) Is the increase in concentration of the lead added to the analysis cell:

100.0 ppb,
or 10.0 ppb,
or 9.1 ppb?

(*ii*) Calculate the concentration of lead in ppb in the sample of drinking water.

SAQ 7.1b

Disadvantages

Although the method involving the addition of a single standard is inherently very rapid, it does suffer from a number of disadvantages.

(*a*) If the analyst makes an error, either during the addition of the standard or when making a physical measurement, there is no in-built check to show that a mistake has been made. The analyst must therefore take great care with all procedures involving the use of this technique.

(*b*) There is no mechanism to average the indeterminate errors that normally occur with physical methods of analysis.

(*c*) Where a sample contains a species that masquerades as the analyte, there is no way for this to be detected and allowed for.

These disadvantages can however be overcome by the use of the *multiple standard-addition technique*.

7.2. METHOD OF MULTIPLE STANDARD-ADDITIONS

In this method we make a number of additions of a standard to the sample solution containing the analyte, and measure the resultant increase in the physical parameter after each addition. The results can best be utilised graphically and the mathematical basis of the graph is given below.

In the previous sections we developed Eq. 7.5 which, you will remember, related to the use of a single standard addition in which no significant dilution occurred.

$$C = \frac{Y_o C_s}{(Y_i - Y_o)} \qquad (7.5)$$

This can be written as,

$$C = \frac{Y_o C_s}{\Delta Y}$$

where $\Delta Y = (Y_i - Y_o)$

and then rearranged to give

$$\Delta Y = \frac{Y_o C_s}{C} \qquad (7.10)$$

Eq. 7.10 shows that there is a linear relationship between ΔY and C_s, the slope of the line being Y_o/C. This is illustrated in Fig. 7.2a.

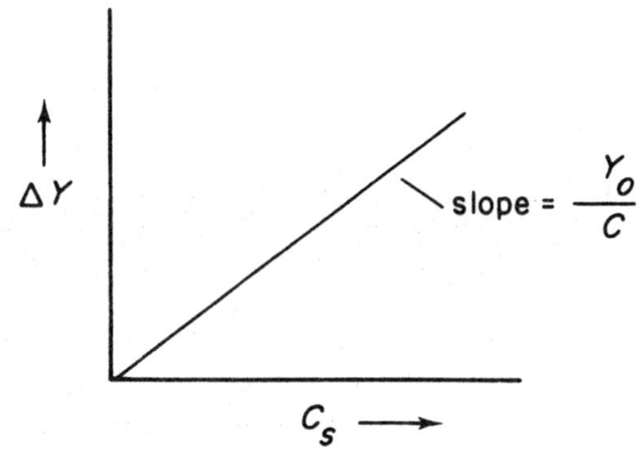

Fig. 7.2a. *Linear relationship of ΔY and C_s (see text)*

As we are now considering that measurements of Y are made after a number of additions of the standard, ΔY is now written as:

$$\Delta Y = (Y_n - Y_o) \qquad n = 1,2,3 \text{ etc},$$

where Y_n is the value of Y after addition of the nth standard.

By measuring the slope of the above line, and knowing the value of Y_o, we can calculate C.

Although this method is quite acceptable, we can in fact use an alternative graphical approach which avoids the necessity of measuring the slope of the line.

We can express Eq. 7.5 in a more general form, *viz*

$$C = \frac{Y_o C_{sn}}{(Y_n - Y_o)} \qquad (7.11)$$

where C_{sn} is the total increase in concentration of the analyte after n additions have been made, and then rearrange Eq. 7.11 to give

$$Y_n - Y_o = \frac{Y_o C_{sn}}{C}$$

Adding Y_o to both sides gives

$$Y_n = \frac{Y_o C_{sn}}{C} + Y_o \qquad (7.12)$$

Eq 7.12 is also that of a straight line, but this time has the intercept Y_o

If we now take Eq. 7.12 and let $Y_n = 0$, we have

$$\frac{Y_o C_{sn}}{C} = -Y_o$$

Dividing both sides by Y_o gives

$$\frac{C_{sn}}{C} = -1$$

or

$$C = -C_{sn}$$

This means, that by constructing a graph of Eq. 7.12 and extrapolating the straight line back to $Y_n = 0$ we have an immediate measure of the value of C. This is illustrated in Fig. 7.2b.

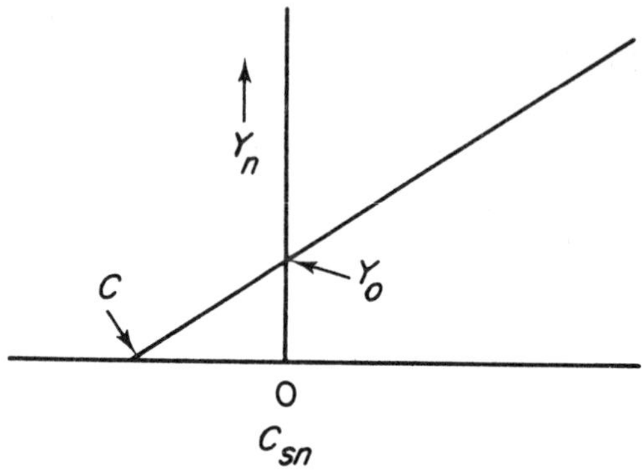

Fig. 7.2b. *Graph of Eq. 7.12 extrapolated*

Let us now compare these two graphical methods by considering an example.

10.0 cm^3 aliquots of a raw-water sample were pipetted into 50.0 cm^3 volumetric flasks. Then 0, 5.00, 10.00, 15.00 and 20.00 cm^3 respectively of a standard solution containing 10.0 ppm of Fe^{3+} were added to the flasks, followed by an excess of aqueous potassium thiocyanate in order to produce the red iron–thiocyanate complex. All the resultant solutions were diluted to volume and the absorbance of each solution was measured at the same wavelength. The results obtained are given in the table below.

Volume of standard added (cm^3)	Absorbance (A)
0	0.215
5.00	0.424
10.00	0.625
15.00	0.836
20.00	1.040

Calculate the concentration in ppm of Fe^{3+} in the raw water sample.

We must first calculate the value of C_s in each of the solutions prepared.

Vol of std added (cm^3)	Mass of std added (μg)	C_s (ppm)	Absorbance (A)
0	0	0	0.215
5.00	50	1.00	0.424
10.00	100	2.00	0.625
15.00	150	3.00	0.836
20.00	200	4.00	1.040

We can now plot absorbance against C_s

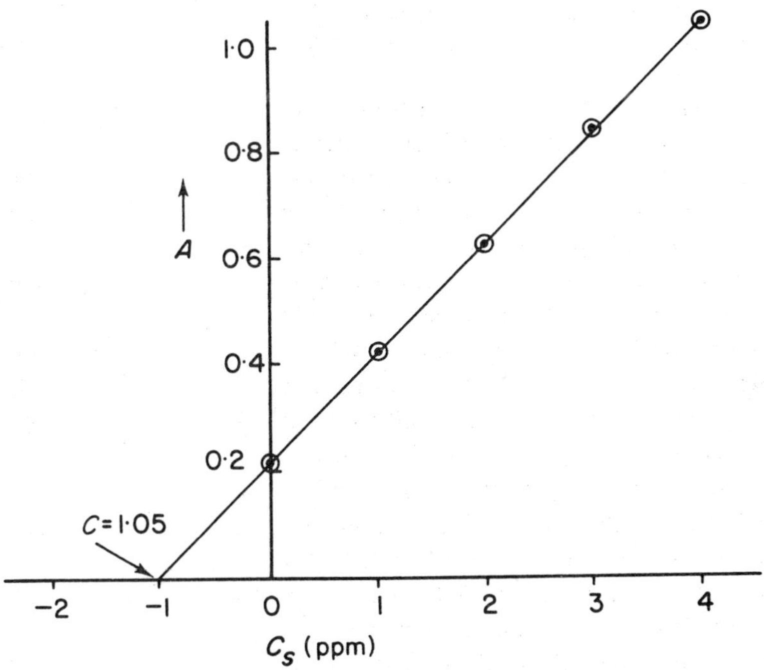

Fig. 7.2c. *Plot of absorbance against C_s (see text)*

From the graph we find that $C = 1.05$ ppm in the solution analysed. However taking into account the dilution factor (10.0 cm^3 diluted to 50.0 cm^3) the actual concentration of Fe^{3+} in the raw water is:

$$1.05 \times \frac{50.0}{10.0} = 5.25 \text{ ppm}$$

Let us now consider the alternative approach to the analysis of the results, by plotting ΔA versus C_{sn}.

C_s	Absorbance (A)	ΔA ($Y_n - Y_o$)
0	0.215	0
1.00	0.424	0.209
2.00	0.625	0.410
3.00	0.836	0.621
4.00	1.040	0.825

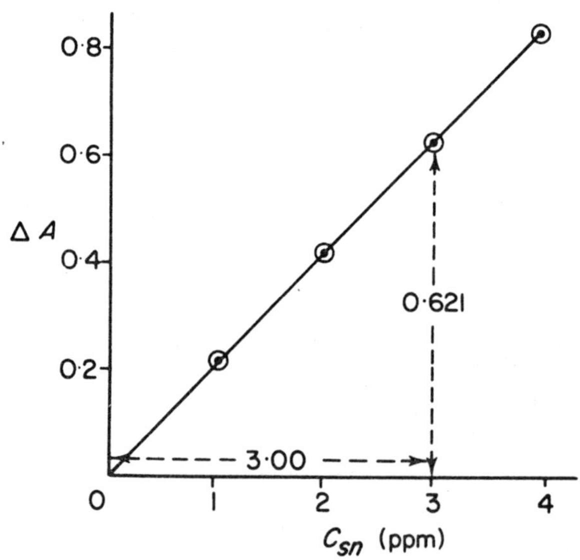

Fig. 7.2d. *Plot of ΔA against C_{sn}*

The slope of the straight line is given by

$$\frac{\Delta A}{C_{sn}} = \frac{0.621}{3.00} = 0.207$$

From Eq. (7.10)

$$\text{slope} = \frac{Y_o}{C} = \frac{A_o}{C}$$

$$\therefore \qquad C = \frac{A_o}{\text{slope}} = \frac{0.215}{0.207} = 1.04 \text{ ppm}$$

Allowing for the initial five fold dilution, the actual concentration of Fe^{3+} in the raw-water is

$$1.04 \times \frac{50.0}{10.0} = 5.20 \text{ ppm}$$

As stressed in Section 7.1 for the single standard-addition method, we must also take into account in the multiple standard-addition approach, significant dilution factors as and when they occur.

Eq. 7.9 relates to a single standard-addition where significant dilution occurs.

$$C = \frac{Y_o C_s(V + v)}{Y_i(V + v) - Y_o V} \tag{7.9}$$

If we let $Y_i = Y_n$ and $C_s = C_{sn}$, Eq. 7.9 can be rewritten as follows.

$$C = \frac{Y_o C_{sn}(V + v)}{Y_n(V + v) - Y_o V} \tag{7.13}$$

Rearrangement gives

$$Y_n(V + v) - Y_o V = \frac{Y_o C_{sn}(V + v)}{C}$$

By adding $Y_o V$ to each side and then dividing by V we get

$$Y_n \frac{(V + v)}{V} = \frac{Y_o C_{sn}(V + v)}{CV} + Y_o \qquad (7.14)$$

If we now plot a graph of

$$Y_n \frac{(V + v)}{V}$$

against

$$\frac{C_{sn}(V + v)}{V}$$

we obtain a straight line with intercept Y_o. If we extrapolate the line to the point where

$$Yn \frac{(V + v)}{V} = 0$$

then,

$$\frac{Y_o C_{sn}(V + v)}{CV} = -Y_o$$

Dividing both sides by Y_o gives

$$\frac{C_{sn}(V + v)}{CV} = -1$$

or

$$C = -C_{sn} \frac{(V + v)}{V}$$

Now at the concentration of the analyte related to the initial volume V (ie $v = 0$)

$$C = -C_{sn}$$

The graph is shown in Fig. 7.2e

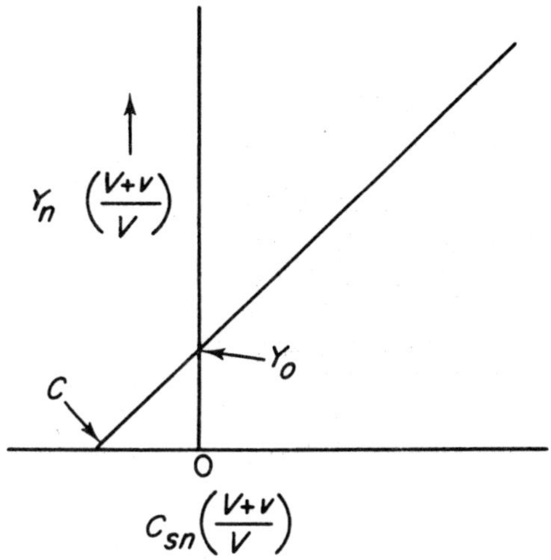

Fig. 7.2e. *Plot of* $Y_n \frac{(V + v)}{V}$ *against* $C_{sn} \frac{(V + v)}{V}$

We can divide both sides of Eq. 7.14 by the factor

$$\frac{(V + v)}{V}$$

We should then be able to plot Y_n against C_{sn}.

Why is this not feasible in terms of obtaining an accurate value for C?

If we rearrange Eq. (7.14) by dividing both sides by

$$\frac{(V + v)}{V}$$

we obtain

$$Y_n = \frac{Y_o C_{sn}}{C} + \frac{Y_o V}{(V + v)}$$

where the term

$$\frac{Y_o V}{(V + v)}$$

is no longer a constant, but decreases with increasing dilution. Thus the straight line relationship is no longer obeyed.

Now as you can see, the direct use of Eq. 7.14 makes for a rather complex procedure. We can, however, simplify the mathematical process by looking more closely at what is being achieved by multiplying the measured parameter Y_n and the concentration of the standard C_s by

$$\frac{(V + v)}{V}$$

Essentially we are

(*a*) correcting the value of Y_n for dilution effects,

(*b*) adjusting the concentration C_s to that operating had a dilution effect not occurred.

For the concentration term therefore, we are *now assuming that the volume remains constant*. We can therefore *replace the concentration axis by a mass axis*, and plot corrected Y_n values against the mass of the standard added. The intercept on the mass axis is now the mass of the analyte in the sample analysed.

SAQ 7.2a

In the polarographic determination of Cd^{2+} in an aqueous solution the following procedure was adopted. 25.0 cm^3 of a sample solution containing an unknown quantity of Cd^{2+} was pipetted into a polarographic cell and the diffusion current (i_d) was measured. To this solution was then added 1.0 cm^3 aliquots of a standard Cd^{2+}

→

SAQ 7.2a (cont.)

solution containing 100 ppm of Cd^{2+}, the diffusion current being measured after each addition of the standard solution. The results obtained are given in the table below.

Diffusion current, i_d (μA)	Volume of std added (cm^3)
10.6	0
17.1	1.0
23.1	2.0
23.7	3.0
34.0	4.0

Assuming that all measurements have already taken into account any blank value, calculate the concentration of Cd^{2+} in the sample solution.

7.2.1. Advantages of the Multiple Standard-addition Procedure over the Single-standard Procedure

(*a*) The normal indeterminate error present in results obtained from any physical method of analysis is averaged by graphical treatment. If necessary, we can use the least squares method to define the best straight line through the points obtained.

(*b*) Operator error can sometimes be detected by a significant change in slope or position of the graph.

(*c*) The presence of an interfering species masquerading as the analyte can sometimes be identified.

SAQ 7.2b

The three graphs (A), (B) and (C), shown below all relate to the same analysis performed by using the multiple standard-addition procedure. Can you identify, if and when an error has occurred in the graphs, and suggest possible sources of that error?

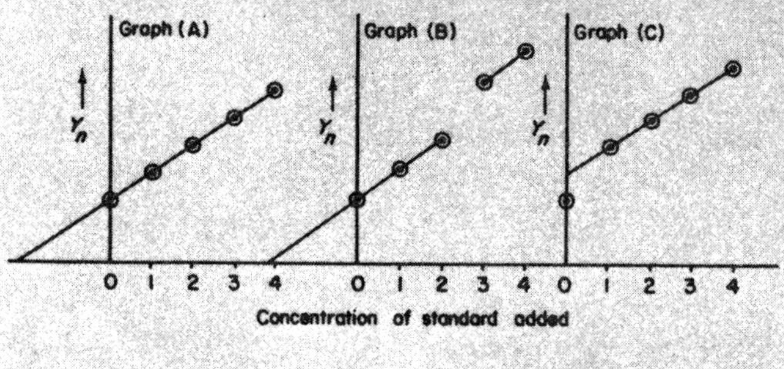

SAQ 7.2b

7.3. OPTIMUM NUMBER AND VALUE OF STANDARDS

Please refer back to all the examples of standard addition that have been quoted in previous sections. You will have noted that an attempt has been made in each example to increase the concentration in the analysed solution by about 100% after the addition of the first or single standard. In the multiple standard-addition approach, it is usual to make each subsequent addition the same as the first. The procedure requires a knowledge of the approximate analyte concentration or an initial trial-and-error investigation before the analysis has to be carried out. As to the number of standards to be recommended for use with the multiple standard-addition method, two factors must be taken into account:

(*a*) the extent of linearity in the analytical method being used;

(*b*) the precision required from the analysis. (This is often related to the monetary value of the product being analysed).

Now it is obvious that we can draw a perfect straight line through two data-points, but a much more representative one through twenty. We thus have to compromise, bearing in mind the two factors stated above. In practice it is normal to collect *five data-points*, that for the analyte and four standard additions.

SAQ 7.3a

A sample of water known to contain sodium and potassium salts is to be analysed for its sodium content by atomic absorption spectroscopy. When this technique is used, potassium is known to interfere with the analysis by enhancing the sodium absorbance. If the sample is thought to contain between 100 and 150 ppm of sodium, suggest an optimum procedure for the determination by using a single standard-addition technique.

The calibration data obtained from a previous experiment for the determination of sodium in the presence of potassium are given in the table below.

Concentration of sodium (ppm)	Absorbance (A)
1.0	0.170
2.0	0.335
3.0	0.505
4.0	0.640
5.0	0.750

SAQ 7.3a

7.3.1. Comparision of Standard-addition Procedures with Calibration Graphs

Although the graphical calibrations described in section 5.2 of this Unit are by far the most important method of calibration, they suffer from two disadvantages.

(*a*) It is undesirably time consuming to prepare a calibration graph when only one or possibly two samples are to be analysed.

(*b*) Components present in the sample may interfere with the analysis in either an additive or a proportional way. Thus we must take care either to remove the interferent if its effect is additive, or 'matrix-match' the standards with the sample if the interferent exhibits a proportional effect.

The standard-addition method may in fact overcome proportional interference effects (enhancement or reduction of the analytical signal) because the standard will suffer the same interference as the analyte already present in the sample. For example, the presence of aluminium in the determination of calcium by atomic absorption spectroscopy, causes a reduction in the calcium absorbance. If the analysis is carried out by standard addition, the same quantity of aluminium will be present in the samples and standards alike, and we can assume that this proportional interference effect has been allowed for.

Finally standard-addition procedures are particularly recommended for 'one-off' analyses because of the limited number of measurements that must be made. However, we must stress that prior knowledge of the necessary linearity in the system is essential, and where greater precision is required the multiple standard-addition approach should be adopted.

List of Objectives

When you have completed this section you will be able to:

- appreciate the value of the standard-addition technique particularly for 'one-off' analyses;
- develop suitable equations for use with the standard-addition technique;
- calculate analyte concentrations from results obtained *via* a standard-addition approach;
- recognise errors or interference effects that occur during a standard-addition procedure.

8. Internal-standard Methods

In all the calibration procedures so far in this Unit, we have made two tacit assumptions:

(*a*) the quantity of sample taken either for calibration or for analysis was accurately known;

(*b*) all the stages in the procedure are the same and the parameters that affect the overall physical measurement have been held constant.

There are, however, important analytical methods where it is not possible to guarantee either the reproducibility of the procedure or the quantity of sample taken for analysis. The most important analytical techniques where these factors occur are:

(*a*) gas liquid chromatography (glc),

(*b*) arc/spark emission spectroscopy and spectrography,

(*c*) X-ray fluorescence analysis.

The method which is often used to overcome these problems is known as the *method of internal standards*. In the internal-standard method, the detector response given by the analyte is compared with that given by another element or compound of known concentration (the internal standard), which is also present in the sample when the analysis is carried out.

Although the internal standard may already be present in the sample as part of its overall constitution, in the majority of cases when the technique is used, the internal standard is added to the sample before the analysis. We shall consider at a later stage, the properties necessary in a compound or element for it to be chosen as an internal standard. When the internal standard is added to the sample before analysis, a solution of the sample is often involved. This must be taken into account during the final calculation.

The technique of arc/spark emission spectroscopy and spectrography are the major examples in which an internal standard is considered to be present already in the sample to be analysed. In the arc/spark techniques, the metal constituents of a sample give rise to narrow spectral lines. The intensity of these lines is a function of the concentration of the element present in the sample. When samples contain an element of approximately constant concentration, for instance the iron in certain grades of steel, and zinc in certain grades of brass, these elements can be used as the internal standard when we analyse for one or more of the minor elements also present in the sample.

To illustrate how the method works we shall consider an imaginary analytical technique in which the measured parameter for a single analyte obeys a relationship consisting of three terms:

ie
$$Y = mkC \tag{8.1}$$

where Y is the measured parameter,

m is a constant term – effectively the detector response,

k is a variable parameter whose exact value may not be known,

C is the concentration of the analyte.

You will recognise that Eq. (8.1) differs from a simple linear equation, in that it contains an extra term, k. This term k is excluded when calibration-plot or standard-addition techniques are being used, as then the value of k is assumed to be constant.

For example, in quantitative spectroscopy, the relationship between absorbance (measured parameter) and the other terms in the equation is given by the Beer–Lambert law, Eq. 8.2.

$$\text{Absorbance} = \epsilon C l \tag{8.2}$$

where ϵ is a constant term known as absorptivity,

l is the path-length of the sample through which radiation is absorbed,

C is the concentration of the absorbing species in the measured solution.

We can assume that

$$\text{Absorbance} \propto C$$

only when the path length l remains constant. Therefore when constructing a calibration plot or using the method of standard addition, a cell of fixed path-length is used throughout.

For the two compounds that are involved in the analysis (ie analyte and internal standard) we can write separate equations of the form of Eq. 8.1. If we let subscript (1) refer to the analyte and subscript (2) to the internal standard then,

$$Y_1 = m_1 k_1 C_1 \tag{8.3}$$

and

$$Y_2 = m_2 k_2 C_2 \tag{8.4}$$

We now divide Eq. (8.3) by Eq. (8.4) to give Eq. (8.5)

$$\frac{Y_1}{Y_2} = \frac{m_1 k_1 C_1}{m_2 k_2 C_2} \tag{8.5}$$

Now if both measurements Y_1 and Y_2 are obtained on the same portion of the analysis sample, it can often be assumed that when they were obtained, the variable parameters k_1 and k_2 were equal.

Eq. 8.5 is then simplified to give Eq. 1. 8.6

$$\frac{Y_1}{Y_2} = \frac{m_1 C_1}{m_2 C_2} \tag{8.6}$$

Within this equation the factor m_1/m_2 is termed the *response ratio*, and can be denoted by the symbol $R_{m1/m2}$.

Because m_1, and m_2 are themselves constants for a given set of experimental conditions, the ratio between them must also be a constant. Thus Eq. (8.6) may be rewritten as

$$\frac{Y_1}{Y_2} = R\frac{C_1}{C_2} \tag{8.7}$$

where $R = \dfrac{m_1}{m_2}$

This is the fundamental equation for use with the internal-standard method, and can be applied in either of two ways.

8.1. METHOD OF SINGLE INTERNAL-STANDARD

In this method it is first necessary to determine the value of the response ratio (R). This is done by preparing a solution containing known quantities of both the analyte and the internal standard. Analysis of this solution will give the two values Y_1 and Y_2. As we already know C_1 and C_2, $R_{1,2}$ can be calculated.

If we now add a known quantity of the internal standard to a known amount of the sample solution, and again measure the value of Y_1 and of Y_2, the concentration of analyte in the sample can be found from Eq. 8.8.

$$C_1 = \frac{Y_1 C_2}{Y_2 R} \tag{8.8}$$

Π An aqueous effluent containing a small quantity of 2-methoxyethanol was analysed by glc by using the procedure outlined below, with 2-ethoxyethanol as the internal standard.

To a 50.0 cm^3 sample of the effluent known to contain 2-methoxyethanol was added 0.119 g of 2-ethoxyethanol, and the resultant solution was diluted to 100.0 cm^3 in a volumetric flask. The prepared sample was then analysed by glc, and the peak areas due to the two compounds were as follows.

	Peak area (mm^2)
2-Methoxyethanol 2MO	1465
2-Ethoxyethanol 2EO	1798

If the response ratio $R_{2MO/2EO}$ calculated from a previous experiment was 0.945, calculate the concentration of 2-methoxyethanol in the aqueous effluent in g dm^{-3}.

In quantitative glc, the measured parameter is the area under the chromatographic peak (ie peak area). The relationship between the peak area of the analyte and of the internal standard is given below.

$$\frac{\text{Peak area of 2MO}}{\text{Peak area of 2EO}} = R\frac{C_{2MO}}{C_{2EO}}$$

The concentration of the internal standard 2-ethoxyethanol, C_{2EO}, in the solution analysed is calculated as follows.

$$C_{2EO} = \frac{0.119}{100} \text{ g cm}^{-3} = 1.19 \times 10^{-3} \text{ g cm}^{-3}$$

Given that $R = 0.945$,

we can now calculate the concentration of the analyte also present in the solution analysed.

$$C_{2MO} = \frac{C_{2EO} \times \text{Peak area for 2MO}}{R \times \text{Peak area for 2EO}}$$

$$= \frac{0.00119 \times 1465}{0.945 \times 1798}$$

$$= 1.03 \times 10^{-3} \text{ g cm}^{-3}$$

We must remember however that the aqueous effluent was diluted two-fold before analysis, and thus the concentration of the 2-methoxyethanol in the aqueous effluent was

$$2 \times 1.03 \times 10^{-3} \text{ g cm}^{-3} = 2.06 \times 10^{-3} \text{ g cm}^{-3}$$

$$= 2.06 \text{ g dm}^{-3}$$

SAQ 8.1a

In the determination by hplc of the vitamin C and the saccharin content of a soft-drink concentrate, caffeine was used as the internal standard. The procedure used is outlined below.

A standard solution was prepared which contained the following concentrations of the three compounds:

Vitamin C (250 ppm w/v), saccharin (100 ppm w/v) and caffeine (150 ppm w/v).

A portion of the standard solution was analysed by hplc and the resultant chromatogram was displayed by means of a potentiometric chart recorder. The peak heights for the three compounds were measured and the results obtained are listed below. →

SAQ 8.1a (cont.)

Compound	Peak height (cm)
Vitamin C	15.15
Saccharin	10.50
Caffeine	12.25

20.0 cm^3 of a soft-drink concentrate was pipetted into a 100.0 cm^3 volumetric flask, and to it was added 5.00 mg of pure caffeine. The mixture in the flask was then diluted to volume with distilled water. On analysis by hplc under conditions similar to these used for the initial standardisation, the following peak heights were obtained for the three compounds of interest.

Compound	Peak height (cm)
Vitamin C	6.95
Saccharin	5.00
Caffeine	8.30

Assuming that over the concentration range examined, a linear relationship exists between the measured peak height and concentration for all three compounds, calculate the concentration (mg dm^{-3}) of vitamin C and of saccharin in the soft-drink concentrate.

SAQ 8.1a

8.2. METHOD OF MULTIPLE INTERNAL-STANDARDS

In this variation of the method, a series of calibration solutions are prepared containing different concentrations of the pure analyte (C), together with a constant known concentration of the internal standard (C_2).

We can therefore rewrite Eq. 8.7 as

$$\frac{Y_1}{Y_2} = \frac{R \times C}{C_2}$$

$$= R'C_1 \qquad (8.9)$$

where R' is the quotient of the constant response factor (R) and the constant known concentration of the internal standard (C_2).

A graph can therefore be plotted of the ratio of Y_1/Y_2 against C_1, which should produce a straight line through the origin with a slope of R'.

If the unknown sample is treated in the same way as the standards used to prepare the calibration plot, the concentration of analyte can be found by simple interpolation from the calibration plot.

Π The ethanol content of a sample of commercial beer was determined by using glc with propan-1-ol as the internal standard.

The following procedure was adopted. To each of a series of six 100.0 cm^3 volumetric flasks was added 3.00 cm^3 of propan-l-ol. To five of the flasks were then added respectively 1.00, 2.00, 3.00, 4.00 and 5.00 cm^3 of pure ethanol, and the resultant solutions were then diluted to volume with distilled water. The propan-l-ol in the sixth flask was diluted to volume with the beer sample to be analysed. All six solutions were then analysed by glc, the areas underneath the ethanol and the propanol peaks being recorded. The results obtained are given in the table below.

% Ethanol (v/v)	Area of ethanol peak (mm^2)	Area of propan-l-ol peak (mm^2)
1.00	1500	4685
2.00	2817	4410
3.00	4715	4900
4.00	6020	4700
5.00	7000	4350
Beer sample	5516	4500

Calculate the percentage by volume of ethanol in the beer sample.

First of all we need to find the ratio of the peak areas of the two alcohols present in the solutions analysed.

% Ethanol	$Area_{EtOH}$ (mm^2)	$Area_{PrOH}$ (mm^2)	Ratio $Area_{EtOH}/Area_{PrOH}$
1.00	1500	4685	0.320
2.00	2817	4410	0.639
3.00	4715	4900	0.962
4.00	6020	4700	1.281
5.00	7000	4350	1.609
Beer sample	5516	4500	1.226

We can now construct a calibration plot of the ratio of the peak areas vs % ethanol (v/v)

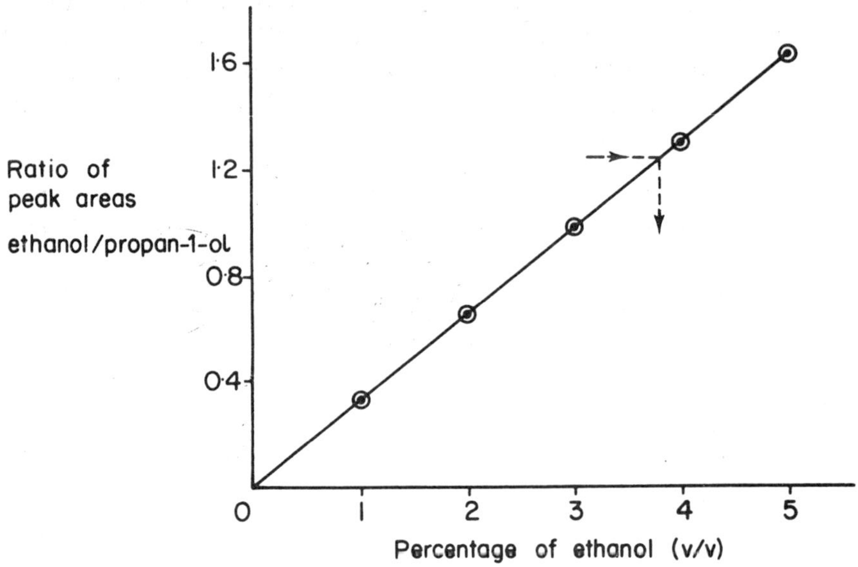

Fig. 8.2a. *Calibration plot*

From the graph the measured ratio of 1.226 ≡ 3.80% ethanol. However we must not forget that the beer sample was diluted by a small amount before the analysis,

ie 97.0 cm^3 of beer were diluted with 3.0 cm^3 of propan-l-ol. Therefore the % of ethanol found must be multiplied by 100.0/97.0 to obtain the true percentage of ethanol in the sample analysed.

$$\therefore \quad \% \text{ ethanol in beer} = 3.80 \times \frac{100.0}{97.0}$$

$$= 3.92\% \text{ (v/v)}$$

In the two previous examples we have used glc to illustrate the application of the internal-standard method, as this represents the major application of this quantitative method.

If you look at the areas underneath the propan-l-ol, peaks in the table you will see a variation in their values. Given that all of the solutions contained 3.0% v/v of propan-l-ol – why this variation?

If we were in fact capable of introducing manually, accurately reproducible volumes of sample for glc analysis, then these areas would have been considerably closer together than was observed. However when handling the very small quantities of sample used in glc analysis, accurate reproducibility of sample volumes becomes virtually impossible without the use of some form of auto-injector. The internal standard method is therefore used to overcome the variation in this parameter, the volume of sample analysed.

8.2.1. Comparison of the Single and the Multiple Internal-standard(s) Method

We can compare the two ways in which the internal-standard method is used in terms of speed of analysis, reliability of the results obtained, and compliance with linearity between the ratio of the measured parameters Y_1/Y_2 and concentration of the analyte.

(a) Speed of analysis

The single internal standard method offers an extremely rapid method of analysis. Once the value of the response factor has been determined, then theoretically this value can be used for any number of similar analyses. In practice the value may alter slightly with changing experimental conditions, and therefore it is normal to check, and update regularly the value of the response factor.

The single internal-standard method is often applied when a number of species are to be determined in a single sample. A single internal-standard can be used for all the species to be determined, provided individual response factors have previously been measured.

(*b*) *Reliability of results*

We have shown earlier in this Unit, that any quantitative method which is based upon a single calibration point must be inherently less reliable than a method based upon a set of calibration points. The data used to construct the calibration allow the individual indeterminate errors associated with each measurement to be evened out. This increases the precision of the analytical results obtained.

(*c*) *Dependence upon linearity between the ratio of the measured parameters Y_1/Y_2 and the analyte concentration*

As in any application of the internal-standard method, the *quantity of the internal standard present is known* and the general Eq. 8.7

$$\frac{Y_1}{Y_2} = R\frac{C_1}{C_2} \tag{8.7}$$

can always be reduced to Eq. 8.9

$$\frac{Y_1}{Y_2} = R'C_1 \tag{8.9}$$

by assuming that the response ratio (R) remains constant. *This assumption is inherent, when we use the single internal-standard method.*

However when we use the multiple internal-standard approach, we are effectively plotting a form of calibration graph, and as with other calibration methods it is *not essential that the graph be a straight line*. When a curved plot is produced, the value of the response ratio (R) is not constant over the whole of the concentration range being examined. However it does not exclude the use of the internal standard method provided that the curved calibration graph is reproducible and is accurately drawn.

SAQ 8.2a

When using

(*A*) the single internal-standard method,

(*B*) the multiple internal-standard method,

state which of the following are assumed to be constant:

(*i*) The ratio of the measured physical parameters,

(*ii*) The ratio of the detector response given by the analyte and by the internal standard,

(*iii*) The concentration of the analyte,

(*iv*) The concentration of the internal standard.

8.3. CHOICE OF AN INTERNAL STANDARD

There are a number of factors to be borne in mind when choosing a suitable internal standard

Π Look back at the two previous in text questions. Can you suggest some factors that are taken into account for selecting an internal standard for use in glc?

In both of the questions the internal standard chosen was *a homologue of the analyte* and therefore is likely to have *similar chemical and physical properties*. It is also likely *to be separable from the analyte* under optimum experimental conditions.

There are always several specific factors which relate to the use of the internal-standard method for a particular analytical technique. These will be considered and discussed in other Units of this course. Let us however consider the general factors, appropriate to all applications of the internal-standard method that affect the choice of a suitable standard.

The internal standard chosen:

(*a*) must not already be present in the sample to be analysed;

(*b*) must be separable from the analyte(s), and all other components present in the sample;

(*c*) must not react chemically with the sample, or interfere in any way with the analysis;

(*d*) must be miscible with the solvent when analyses of solutions are being carried out.

SAQ 8.3a

Select from the following list of compounds, those which you consider *might be useful* as internal standards for the glc analysis of an aqueous effluent containing traces of butan-2-one ($C_2H_5COCH_3$).

Hexane	**(C_6H_{14})**
Pentan-2-one	**($CH_3CH_2CH_2COCH_3$)**
Propanone	**(CH_3COCH_3)**
Toluene	**($C_6H_5CH_3$)**
Acetyl chloride (ethanoyl chloride)	**(CH_3COCl)**
Propan-2-ol	**($CH_3CH(OH)CH_3$)**

8.3.1. Main Advantage of the Internal-standard Method

The main advantage offered by the internal standard method over other calibration procedures so far described is that it is capable of providing a result of somewhat higher accuracy and precision than otherwise. Under the best conditions, an overall accuracy of $\pm 0.5\%$ of the true value, can be achieved for an instrumental analytical method.

Remember that for instrumental methods, and results obtained by the usual calibration procedures, an accuracy of ± 1–2% is normally acceptable. The higher accuracy and precision obtained by using the internal-standard method, is considered to be due to the avoidance of some of the factors which lead to overall inaccuracy of a result. This is achieved because the concentration of the analyte and that of the standard are both measured;

(*a*) in the same portion of sample solution,

(*b*) either at the same time or at times very close to one another.

The use of the internal-standard method is therefore not restricted to the analytical techniques described earlier in this section, but can be usefully employed in other techniques when increases in accuracy and precision are required.

List of Objectives

When you have completed this section you will be able to:

- define what is meant by an internal standard, and to formulate the mathematical relationship upon which the required calculation is based;

- appreciate how and where the internal-standard method may be applied;

- calculate analyte concentrations by using the internal-standard method;

- choose suitable internal standards;

- appreciate the advantages of the internal-standard method over other calibration procedures.

9. Calibration by Computational Means

Microcomputers and microprocessors have become an integral part of many modern laboratory instruments. They can perform a number of functions, some of which are discussed in 9.1 below.

9.1. FUNCTIONS PERFORMED BY A MICROCOMPUTER IN MODERN LABORATORY INSTRUMENTATION

The interaction between a computer and an analytical instrument may be classed as either *active* interaction or *passive* interaction. Most analytical laboratories use computers mainly in the *passive* mode ie to acquire, manipulate, and store data. The *active* mode is more often found in chemical plants where total automation of the analysis is required, ie sampling through to calculation.

In most analytical laboratories where computers have been introduced, the small and often dedicated microcomputer is used, as opposed to a traditional 'main frame' system. Therefore the term 'computer' as used hereafter refers to a 'microcomputer' or a 'microprocessor'. Let us now consider in more detail the active and passive applications of computers in laboratory instrumentation. We shall concentrate on the role they play in calibration.

9.1.1. Passive Applications

Most of the uses to which we put computers in a modern analytical laboratory come under the heading of passive applications. The main uses are discussed below.

Data Acquisition

The results arising from an analytical instrument can either be recorded manually or passed directly to a computer. Manual recording may be subject to human error when we transpose the results from the instrument to our analytical notebook. The direct acquisition of data by the computer poses no such problem. The computer can also average data rapidly, thereby giving a more accurate value for any given measurement.

Data Processing

We have already seen in this Unit that calibration procedures all required relatively simple mathematical manipulation of the results acquired. In constructing calibration plots, the computer does not possess the artistic judgement of the human eye and therefore constructs the graph by using traditional mathematical principles. This is a more precise method than that often adopted by the analytical chemist.

For example, if we know that a calibration graph is linear over a given range of concentration, then when programmed to do so, the computer will automatically carry out a linear regression analysis on the data acquired, and thereby construct the best straight line through these data-points. If on the other hand we know that the data may lie on a smooth curve, then the computer can be programmed to generate a polynomial expression which best fits the data.

A polynomial expression refers to an equation giving say y in terms of x, where x is present in a number of powers.

eg $$y = ax^3 + bx^2 + cx + d$$

and where *a*, *b*, *c* and *d* are constants.

For a linear equation as we have seen, the polynomial is simplified to that involving *y* and the first power of *x*.

ie $$y = ax + b$$

For a smooth curve, a quadratic equation of the type

$$y = ax^2 + bx + c$$

may be required to describe the curve.

Data Storage

The computer can store vast quantities of data. These data are generally stored on magnetic discs and transferred to the central processing unit of the computer as and when required. If the analysis is to be carried out by using calibration data obtained at an earlier date and held in store by the computer, then it is essential that we check the current validity of the data by using one or more accurately prepared standards. Given that the analytical method to be used is subject to slight instrument variation, then the computer programme may be capable of applying a factor to the calibration data, which allows for the instrument changes that have occurred. This is sometimes termed 're-sloping' of the calibration plot. This process of 're-sloping' should be applied only when you have attempted exactly to reproduce experimental and instrumental conditions, otherwise a new calibration graph should be constructed

Communication of Results

In all modern computer systems, the data which are communicated between the instrument and the computer are displayed on a visual display unit (VDU) monitor. This enables us:

(*a*) to check that the correct analysis conditions have been selected;

(*b*) to check the validity of the input data eg calibration concentration, units;

(*c*) to make an immediate assessment of the precision of the calibration data;

(*d*) to obtain an immediate result in concentration terms once the analysis has been carried out;

(*e*) to evaluate the result in statistical terms.

Assuming that the computer has printer capabilities either built-in or attached, then the results obtained from our analysis, together with the calibration, data used to obtain them can be readily 'output' in a suitable form for inclusion within an analytical note book. This procedure forms part of what is generally now termed *good laboratory practice* (glp).

Good laboratory practice is a general term relating to a set of rules, operating procedures, and practices that have been established by a regulatory authority, and are adequate to ensure the quality and integrity of the data generated by a laboratory.

9.1.2. Active Applications

As we have stated earlier in this section, the use of computers in the 'active' mode is mainly found in chemical plants where analysis is often carried out on a totally automated basis. In the laboratory however 'active' applications are usually restricted to the control of instrumental operating conditions.

For example, these may be the adjustment of:

(*a*) the slit width and the wavelength setting of a monochromator,

(*b*) the temperature of a chromatographic column and the time at which the integration of a chromatographic peak is to begin, (*c*) the rate of addition of a reagent in an automatic titration, and the prediction of the end point in that titration.

The examples quoted are all rather simple and obviously do not reflect the total capability of the modern microcomputer. A much more detailed survey of computers in analytical chemistry can be found in the Unit of this course dealing with the application of microprocessors.

9.2. CALIBRATION TECHNIQUES WITH COMPUTER-CONTROLLED OR COMPUTER-ASSISTED INSTRUMENTATION

Dependent upon the technique and the sophistication of the instrumentation, calibration can be carried out either manually or automatically.

9.2.1. Manual Calibration

By manual calibration, we mean that the calibration standard is introduced into the instrument at a time decided by the operator. This standard may well be used either to check the current validity of the instrument calibration or automatically to update the calibration already held by the computer. For routine analysis we should probably check the calibration once every 10 to 20 samples, but the frequency of checking may well be increased when a high level of accuracy is required.

9.2.2. Automatic Calibration

Some of the more sophisticated laboratory instrumentation, and certainly automatic equipment for use in on-line analysis, can automatically check and update the calibration. These instruments will sample automatically, and are programmed to take a sample of standard at either a fixed time or a fixed sample-interval. The computer updates its calibration information each time a standard is taken and all samples which follow the standard are automatically related to it.

For this single-point calibration to be really effective however, the relationship between the measured parameter and the analyte concentration should preferably be linear. Also, from previous experience, we should know that the calibration changes only very slightly over long periods of time. A number of factors must be taken into account when this procedure or method is to be adopted.

(*a*) The standard solution must be stable over long periods of time.

(*b*) Intermixing of samples and standards must be avoided.

(*c*) The instrument should be capable of providing reproducible analytical conditions for both calibration and analysis samples (eg control of temperature).

(*d*) The instrument should have a warning device which is activated when two successive standards differ by more than an acceptable amount.

9.3. THE LABORATORY DATA STATION

With the power and capabilities of microcomputers now available, some laboratories, particularly those engaged in routine or quality control analysis, are moving towards a central laboratory microcomputer, termed a *laboratory data station*, rather than a host of smaller computers dedicated to individual instruments. The instruments within the laboratory are connected *via* cable to the data station, which holds on file all the programmes relating to the attached instruments. Depending upon the sophistication of the laboratory data station, it may well have the only keyboard for manual input of data and the only VDU within the laboratory. Alternatively, a VDU and keyboard may be associated with every instrument attached to the data station.

9.4. EXAMPLE OF LABORATORY BASED COMPUTERISED INSTRUMENTATION

We can purchase microprocessor-attached or computerised instrumentation for every analytical technique commonly in use in the laboratory. Let us look in detail at one particular technique and consider what functions the computer performs when an analysis is carried out with this equipment. The technique we shall use for illustration is atomic absorption spectroscopy, the apparatus consisting of three items as shown in Fig. 9.4a.

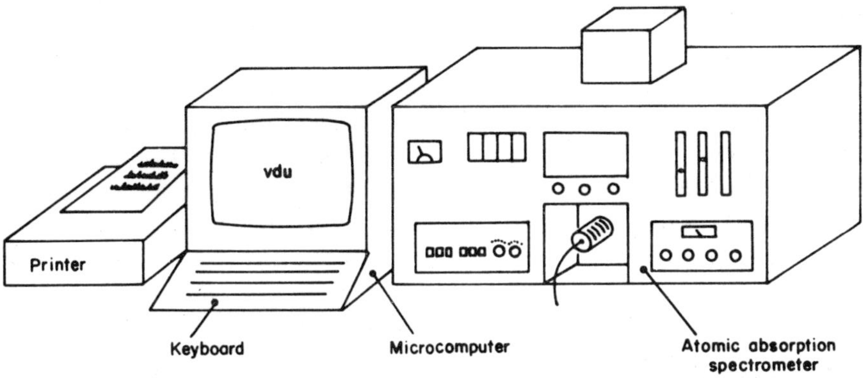

Fig. 9.4a. *Schematic diagram of a modern atomic absorption spectrometer*

We can divide the application of this instrumental technique into the functions carried out by the computer and these carried out by the operator.

9.4.1. Functions of the Microcomputer in this Instrumentation

Data Storage

(*a*) The computer stores method data and instrumental data for the analysis of all metallic elements normally analysed by this technique.

(*b*) The computer stores calibration data obtained by using the conditions given in (*a*) above.

(*c*) The computer stores results of the analyses carried out under the analysis title.

All the above data are held on magnetic discs and transferred to the computer as and when required.

Data Acquisition

The computer is attached directly to the spectrometer to acquire the absorbance data. The data are read far more rapidly than is possible visually (eg 16 times a second). The operator is informed when a sample is to be introduced into the spectrometer (ie when the computer is ready to accept the data), and when the measurement of any single sample is complete.

Data Processing

The majority of analyses by atomic absorption spectroscopy are carried out by using calibration plots, and the programme is written which allows sample data to be compared with the calibration data. The data processing part of the computer therefore performs a number of functions.

(*a*) It averages out the absorbance measurements acquired for each standard and sample analysed.

(*b*) When the instrument is calibrated for a particular analysis, the computer constructs the best line through the data points acquired and may indicate any of the points which look suspect.

(*c*) It displays the calibration data in both tabular and graphical form, which allows the operator to add additional points or substitute points that look suspect.

(*d*) It uses the calibration data for the analysis of samples of unknown composition, and displays the answer in terms of absorbance and concentration.

(*e*) It allows for regular checks on the calibration to be carried out by using a single standard, and applies to the subsequent analysis an appropriate factor, which allows for the slight change in slope that has occurred. If the standard value differs markedly from the original calibration value at that concentration, the computer registers that there is a fault with either the new standard or with the instrumental conditions.

Data Presentation

The computer transfers all the calibration data, graphical plots and results of analysis to a printer. This printed sheet can be stored in the operator's analytical note book.

9.4.2. Functions of the Operator

Preparation of Standards

The operator initially needs to decide whether a new set of calibration data has to be obtained or whether the set of data, already held in file by the computer is satisfactory. Provided the instrumental conditions under which the original data were obtained can be accurately reproduced, it should be possible to use these data for an analysis. If fresh standards are to be prepared then this must be done bearing in mind all that has been explained so far about the preparation of standard solutions.

Setting up the Instrument

The computer details the optimum conditions to be employed, and the operator must adjust the various parameters to these optimum conditions. When the computer exercises total or partial control over these instrumental conditions, the operator's task is clearly reduced. When the operator decides to change the analytical/instrumental conditions, this information must be input into the computer, probably detailed under a new program title.

Performing a Calibration

Assuming that the operator has chosen to produce a new set of calibration data, then he must input into the computer details of the number and concentration of standards, and decide how many absorbance values must be taken before an average value is calculated.

Presentation of Solutions to the Spectrometer

Under instructions from the computer, the operator will present in order, each of the standards to the spectrometer. When the calibration is complete the operator must assess the data and accept or reject them as necessary. Having accepted the calibration data, the analysis of unknowns can be carried out.

Analysis of Unknowns

Again, under instructions from the computer, the operator will present the samples to the spectrometer in a logical order and either instruct the computer how many samples are to be analysed or when the analysis is complete. The average of the absorbance data for the samples should match that of the standards in order to give both the same precision.

Recording of all Data Supplied

At no stage is there any need for the operator to record in writing any of the absorbance measurements obtained. The final print-out can be pasted directly into the operator's notebook, and he can choose whether to store the data aquired on magnetic disc, or to erase it from the computer's memory.

Π What do you consider to be the advantages offered by the use of the microcomputer in the example detailed above? Assume that the spectrometer is not controlled in any way by the computer.

In considering the advantages offered by the effective addition of a microcomputer to an existing spectrometer, we must state how use of the computer benefits the operator. Let us first list the possible advantages.

(*a*) When the instrument is initially set up for a particular analyte, the operator must check the optimum settings. The computer, via its data package, offers a means by which this can be done rapidly.

(*b*) The averaging of each group of data will increase the overall accuracy of each measurement.

(*c*) The automatic recording of absorbance data eliminates any human error which may occur in manual transposition of data.

(*d*) Once the calibration data have been assimilated by the computer, it derives an equation offering the 'best fit' for the data-points, and plots the resultant graph on the VDU. The operator therefore gets an immediate assessment of the precision of the data. If the operator considers that one of the data-points looks rather imprecise compared to the rest, often it can be eliminated and a new equation derived. Alternatively, and more likely, the operator will prepare a fresh standard at that concentration and substitute the new value for the original one.

(*e*) The concentration of the analyte in the sample analysed is presented directly to the operator.

(*f*) The operator has a permanent record of the analyses carried out and the calibration on which the results were based.

Summarising therefore, the major overall advantages offered by the use of the computerised instrumentation are:

(*a*) an increase in the accuracy of the analysis;

(*b*) a decrease in the time required to obtain the analyte concentration. This represents a distinct advantage when a large number of similar samples are to be analysed.

SAQ 9.4a

The figure below, represents a block diagram of an hplc system interfaced to a microcomputer, which forms a data station within your laboratory.

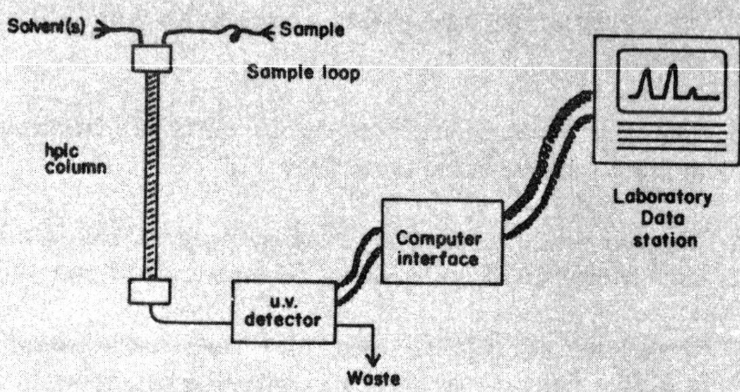

When analysing a mixture containing a number of components by hplc, the individual components eluted from the hplc column are passed into the U.V detector. The resulting absorbance signal from the detector is passed *via* a suitable interface into the microcomputer forming the laboratory data station.

In the mixture to be analysed there is only a single analyte to be determined, and the technique of single standard-addition is to be used. You may assume that no dilution of the analyte occurs when preparing the sample containing the standard added.

(*i*) What information will the data station require to have stored, before the analysis can be carried out and the result presented in concentration terms? ⟶

SAQ 9.4a (cont.)

(*ii*) What information will the operator need to provide to the microcomputer?

(*iii*) What information will the hplc instrument need to provide to the microcomputer?

(*iv*) What functions does the microcomputer have to perform in this analysis?

List of Objectives

When you have completed this section you will:

- know the functions performed by microcomputers in analytical instrumentation;
- appreciate the advantages offered by incorporating microcomputers into analytical instrumentation;
- understand the interaction between the analytical instrument, the microcomputer, and the analyst in the modern analytical laboratory.

10. Comparison of Calibration Procedures

When we choose an appropriate calibration procedure for an analysis we should ask ourselves a number of questions:

— how many samples are to be analysed?

— how experienced is the person carrying out the analysis?

— what do we know about the sample – eg the presence of interfering species?

— what accuracy is required for the analysis?

Let us now attempt to answer each of these questions in turn.

How Many Samples Are to be Analysed?

It is often convenient to divide analyses, in terms of the number of samples, into three categories.

— 'one off' samples,

— analysis of small batches,

— routine analyses.

At the same time we can divide the comparative methods available into those involving

- calibration plots,
- standard-addition,
- internal standards.

The 'one-off' sample is probably best analysed by the single standard-addition procedure. Not only will this automatically compensate for certain interference effects but also requires only two measurements to produce a result. With samples appearing in small batches, a calibration plot should probably be prepared and possibly checked by a single standard before the analysis of each batch. For routine work a calibration-plot procedure should be adopted, and standards used regularly as checks on its continuing validity.

Experience of the Analyst or Technician

Many routine analytical measurements are made in industrial chemical plants, often by technicians who may have had no formal training in analytical chemistry. We should then choose methods which are simple and uncomplicated. Methods involving calibration plots should be used, preferably in conjunction with some computation both to interpret the results and perform the calculation.

Knowledge of the Sample

There are very few analytical methods or techniques that are truly specific for a single analyte, and relatively few analyses are carried out on simple systems. Where analyses are carried out routinely we are likely to be aware of the approximate composition of the sample. Therefore interference effects caused by part of the matrix can be avoided by a suitable choice of analytical method, together if necessary, with an appropriate separation stage, or addition of 'masking' agents. With the 'one-off' analysis however, very little will probably be known about the sample. The method of standard addition is therefore the best choice.

What Accuracy is Required?

When the specification limits for the presence of a component in a matrix are very narrow, or an analysis is required to a high degree of accuracy, then we must inevitably take more care and more time in obtaining the result. Internal standard methods may therefore be adopted. The use of calibration-graph methods with bracketed standards would certainly provide the accuracy, but would probably be chosen only when a very small number of samples were to be analysed.

SAQ 10a

The following list of statements [(i) to (v)], relate to a comparison of two methods of analysis,

(*A*) the single internal-standard method,

(*B*) the method involving the bracketing of standards around the sample.

(*i*) Both these methods require the existence of a linear relationship between analyte concentration and detector response over the range of concentrations examined.

Answer: true, false, or 'insufficient information given'.

(*ii*) Both methods increase the accuracy of a determination.

Answer: true, false, or maybe

(*iii*) Method (*A*) is more rapid to carry out than method (*B*).

Answer: true, false, or 'insufficient information given'. ⟶

SAQ 10a (cont.)

(*iv*) Both methods are capable of overcoming interference effects caused in the analysis by another component in the analyte solution or sample.

Answer: true or false.

(*v*) Both methods are suitable for regular routine analysis.

Answer: true, false, or maybe.

10.1. CALIBRATION METHODS IN GAS ANALYSIS

The difficulty in preparing mixtures of known composition for use as calibration standards increases as we go from the liquid to the solid to the gaseous state. The problems associated with the preparation of gaseous mixtures of accurately known concentrations are due mainly to the normal physical properties of gases. Gases cannot easily be weighed, the volume of a gas may change during handling and we must also take into account temperature and pressure effects. The main result of all of these physical effects is that we cannot in general prepare standards to an accuracy equivalent to that obtainable with either liquid or solid systems.

We can divide the methods for preparing gas mixtures into two categories:

— static methods,

— dynamic methods.

10.1.1. Static Methods

Static methods are those in which a quantity of a standard mixture is prepared in a suitable container before being sampled or passed to the analyser.

We can determine the concentration of the analyte gas or gases in the mixture from measurements made during the mixing or filling process, or alternatively by some form of absolute chemical method after the mixture has been prepared.

Example. Sulphur dioxide will react quantitatively with iodine according to the following equation.

$$SO_2 + I_2 + 2\,H_2O \rightarrow H_2SO_4 + 2\,HI$$

Therefore, by treating a known volume of a gaseous mixture containing sulphur dioxide with an aqueous solution containing an excess of iodine, and back titrating the unreacted iodine with thiosulphate, the quantity of sulphur dioxide in the gas mixture can be determined.

$$[I_2 + 2\,Na_2S_2O_3 \rightarrow Na_2S_4O_6 + 2\,NaI]$$

Suitable vessels for the preparation of gaseous mixtures are metal, glass, or plastic containers. More will be said about this later.

Let us now consider how the mixture of gases can be prepared.

Gaseous Mixture Produced by Pressure

The cylinder or vessel to be filled, is initially evacuated by means of a vacuum pump. Pure analyte gas is then allowed to enter the cylinder *via* a suitable valve system to produce the required pressure. After allowing the cylinder and contents to reach ambient temperature, the diluent gas is added to the cylinder until an appropriate pressure is reached. This may be atmospheric pressure when an immediate supply of the gas mixture is required, or a pressure higher than atmospheric for a permanent supply. The concentration of the analyte gas is calculated by assuming ideal-gas behaviour in the mixture and that Dalton's Law of Partial Pressures is obeyed.

The pressure of an individual gas in a mixture is termed the partial pressure. The total pressure of a mixture of gases is, the sum of the partial pressures of the constituents. This is known as Dalton's Law of Partial Pressures. With pressurised gases we must take into account the compressibility of the two components which will cause a slight deviation from Dalton's Law. A discussion of this issue is however outside the scope of this text.

Gaseous Mixture Produced by Weight

This technique is only really appropriate when pressurised gases are being prepared.

SAQ 10.1a

> **Do you consider, the method of preparing gaseous mixtures by weight would be appropriate to prepare 2 litres of a 10% v/v mixture of CO_2 and nitrogen?**

The container is weighed before and after addition of each component, and the appropriate gravimetric calculation carried out. In order to prepare gaseous mixtures at low levels of concentration by this method, successive dilution of previously prepared mixtures is recommended.

Preparation of a Laboratory Sample of a Gaseous Mixture at Atmospheric Pressure

The average laboratory will not have available facilities for preparing pressurised gases. However we may find it necessary to prepare mixtures of gases or to dilute pressurised gases purchased from a commercial supplier. Such mixtures can be conveniently prepared in glass vessels of known volume or alternatively by mixing known volumes of gases in suitable plastic bags.

Many plastic materials are either permeable to gases or affect the concentration of a mixture by selective adsorption of one of the components. Nylon or Teflon bags have, however, been successfully used. Glass is also capable of selective adsorption of gases onto the walls. Best results are therefore obtained, with both plastic and glass vessels, by preconditioning the vessel with a similar mixture before preparation of the standard.

When plastic bags are to be used the procedure is as follows.

Empty the bag as much as possible by squeezing it or applying a vacuum to it and then half fill it with the diluent gas. Now add the analyte gas and fill the bag with the remainder of the diluent. The components in the mixture are generally added by means of a gas burette or a syringe. The bag is now sealed and shaken to produce a homogenous mixture of gases. This method can also be used to prepare gaseous mixtures from volatile liquids provided the boiling-point of the liquid is about room temperature. The liquid may then be added to the diluent gas *via* a syringe and the quantity added is calculated on the basis that 1 mole of the compound being diluted will occupy 22.4 dm^3 at STP, and that the density of the compound is known. Remember that STP refers to standard temperature and pressure, the temperature being 273 K (0 °C) and the pressure, 1 atm.

SAQ 10.1b

A plastic bag holds 1750 cm^3 of gas at atmospheric pressure. If 100 μl of vinyl bromide (M_r = 116) is introduced into the bag and the pressure of gas in the bag is adjusted to 1.00 atmosphere, calculate the percentage by volume of vinyl bromide in the gaseous atmosphere.

You may assume all the vinyl bromide inside the bag to be vapourised.

Temperature of the gas = 20 °C,
Barometric pressure = 1.00 atmosphere,
Density of vinyl bromide = 0.95 g cm^{-3}.

SAQ 10.1b

10.1.2. Dynamic Methods

Preparation of mixtures by dynamic methods involves the mixing of accurately metered flows of the component gases. Dynamic methods are convenient when preparing a calibration curve which requires several different analyte concentrations or when preparing mixtures having low concentrations of analyte. There are a number of different dynamic methods, some of which are described below.

Simple Flow-mixing

Binary mixtures can be prepared by mixing two streams of known volume flow-rates. These flow-rates are measured by calibrated flowmeters as the gases pass into the mixing chamber. A range of standard mixtures can be prepared by altering the flow-rates of the two streams. The mixing system must be designed to create turbulence in order to homogenise the mixture. The accuracy of the composition of the mixtures prepared in this way, however, will not be very high, as the measrements being made all depend upon the accuracy of the flow controllers and flow-measuring devices.

Gas-mixing with Proportioning Pumps

This is a variation on the mixing of metered flows described above. In this system two reciprocating pumps are coupled by interchangeable gear cogs so that the output produces the required standard mixture. Different mixtures are prepared by changing the gear cogs, and the precise composition of the mixture is a function of the gear-ratio chosen, and the swept volume of the pump cylinder. By using very precise gear wheels, it has been shown that an accuracy of ±0.5% is obtainable. Also, by using two or more pumps in series it is possible to produce standards at concentrations down to ppm levels.

Controlled Injection with Syringe Pumps

Low-concentration standards can also be prepared by injecting a gas or a liquid into a metered flow of diluent gas. A syringe or microsyringe is used to add the gas or the liquid, the rate of movement of the syringe plunger being controlled by a geared-motor drive. As in other methods already described turbulence is needed to ensure efficient mixing and consistent composition.

Use of Permeation Devices

This method is mainly used to prepare mixtures of compounds, the boiling-points of which are below room temperature. The method is based on the fact that plastics are permeable to gases.

The pure liquid is sealed inside a small piece of suitable plastic tubing – Teflon is often used, and then inserted into a stream of diluent gas.

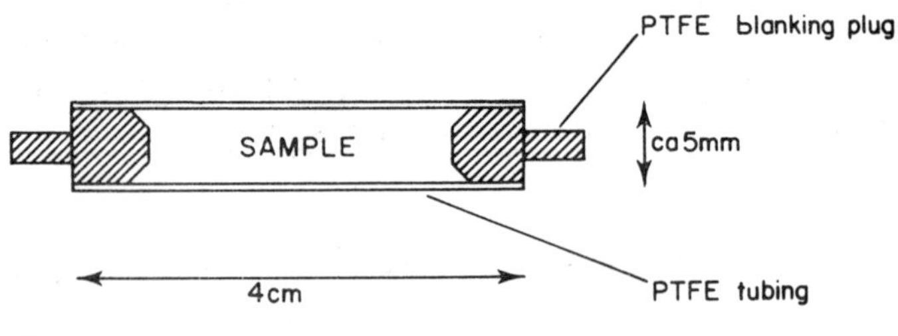

Fig. 10.1a. *A permeation device*

The whole apparatus is thermostatted at a temperature above that of the b.p of the liquid at atmospheric pressure. Under these conditions the rate of permeation through the polymer membrane is a function of :

— membrane area,

— membrane thickness,

— temperature and pressure differences across the membrane.

The diluent gas is allowed to flow at a constant rate over the sealed tube, and the loss in weight of the tube, that occurs during a fixed period of time, allows us to calculate the concentration of the vapourised liquid in the diluent gas. The method is usually employed for preparing mixtures containing small quantities of the analyte.

SAQ 10.1c

Designate 'true' or 'false' the following statements relating to the preparation of gaseous mixtures. If you cannot decide upon an answer, return to the text, as it will be of no value to you to hazard a guess.

(*i*) The gravimetric method of producing gas mixtures is well suited to the preparation of cylinders of mixed gases at high pressures.

True or false?

(*ii*) The preparation of gaseous mixtures by measuring the pressure increase inside an evacuated cylinder, is a good example of a dynamic method.

True or false?

⟶

SAQ 10.1c (cont.)

(*iii*) Most plastic materials are suitable for the preparation and storage of mixtures of gases.

True or false?

(*iv*) Permeation tubes are good devices for the preparation at ambient temperatures, of low-concentration mixtures of relatively low volatility organic compounds.

True or false?

List of Objectives

When you have completed this section you will:

- able to appreciate the factors that must be taken into account when choosing calibration procedures,
- able to judge the appropriateness of a calibration procedure,
- aware of the problems associated with the preparation and storage of gaseous mixtures,
- know some methods for preparing gaseous mixtures.

11. Monitoring the Performance of Analytical Procedures

An analytical laboratory should have, as one of its principal objectives, the production of high-quality analytical results. This can be achieved by the use of accurate and reliable analytical methods, which are appropriate for the analysis being carried out. However, experience has shown that deficiencies may occur when insufficient care is taken, either with the choice of an analytical method or with its application. The growing concern with poor laboratory practices has led to a proliferation of laboratory accreditation programmes, government regulations relating to good laboratory practice, and the development and application of quality control and quality assurance programmes. The common purpose of all of these efforts is to ensure the accuracy, precision, and reliability of analytical results.

The definition of 'good laboratory practice' has been given earlier in this Unit in Section 9.1.1.4. Various definitions have been suggested for quality control and quality assurance. Frequently the terms are used interchangeably.

Quality Control can be defined as 'a planned system of activities, the purpose of which is to secure a product of defined quality.'

Quality assurance can be defined as 'a planned system of activities, the purposes of which is to provide assurance that the quality control programme is really effective.' Every quality assurance programme must have a set of objectives. The objectives will vary from one laboratory to another and will focus largely on the purposes for which the laboratory exists. Now although it is not the purpose of this Unit to consider in detail quality assurance programmes, some of the objectives within any quality assurance programme will relate partly to calibration procedures. Examples are:

— adequate accuracy and precision of data generated by analysts within a laboratory,

— identification of weak methodology,

— ensuring that the analytical work will withstand legal scrutiny in any regulatory action.

Some of the ways in which these objectives can be realised are considered in the sections below.

11.1. USE OF REFERENCE MATERIALS AND CERTIFIED REFERENCE MATERIALS

Definition A material or substance, one or more properties of which are sufficiently well established to be used for calibration purposes, can be called a *Reference Material*. When the value of that property has been certified by a technically valid procedure, and the sample of the material or substance is accompanied by a certificate or other documentation, the material can be defined as a *Certified Reference Material*.

Reference materials need not be pure single substances. They may be in the form of mixed gases, liquids, or solids or even simple manufactured objects. In order to achieve certification the materials may either be analysed individually or be related to representative samples taken from a batch.

The certified value given for the material is often based upon results obtained from a number of laboratories, different methods of analysis being applied. Some of the results obtained for the material analysed may be rejected on the grounds that they differ sufficiently from those obtained from other laboratories. The certified value is thus the mean of the accepted results.

Reference materials are widely used in analytical laboratories:

- to calibrate analytical instruments,

- to validate analytical methods,

- to allow for the transfer of analytical results between laboratories.

Certified reference materials for calibration purposes are normally used when it is difficult for us to produce reliable standards. This may often occur when solid or gaseous matrices are involved. For instance, solid samples are regularly analysed by optical emission spectroscopy, and thus solid standards are required for calibration. Certified reference materials are available not only in a range of matrices, but also containing different quantities of analyte. Before we can transfer analytical results between laboratories, and be confident of the results we are providing, we must be prepared to carry out some check on our analytical method by using an independent standard. A certified reference material of the type already indicated would serve to justify the reliability both of the method and of the results supplied. These are available for a number of sources, two of which are:

- The Bureau of Analytical Standards,

- The Community Bureau of Reference.

SAQ 11.1a

> **What conclusions could you draw, if when analysing a sample of a certified reference material by your own laboratory procedure, you obtained a result different from that of the certified value?**

11.2. USE OF SYNTHETIC SAMPLES AND 'SPIKED' SAMPLES

Both these methods can be used in order to check the validity of an analytical procedure.

A synthetic sample is one which is formulated from known quantities of constituents, so as to represent as closely as possible the 'real' samples to be analysed. We can now analyse the synthetic sample by our procedure and thereby determine the percentage recovery of the analyte. Under normal circumstances, recoveries of between 98 and 102% would be acceptable. The disadvantage of using synthetic samples are:

— the time they take to prepare,

— the knowledge required of the 'real' sample in order accurately to match the essential components.

SAQ 11.2a **How do you consider we arrived at a value of 98 → 102% as an acceptable range for recovery?**

An alternative procedure, which is less time-consuming but still effective is '*spiking*'. There is sometimes confusion in the literature between the terms '*spiking*' and '*standard addition*'. The terms are often used synonymously, the standard added being referred to as the 'spike'. In the context of this part of the Unit, the 'spike' is not added as a means of determining the quantity of analyte, but as a means of validating an analytical procedure.

In this procedure, the concentration of the analyte in the sample, has already been determined or is known to be zero, and the 'spike' (ie a small quantity of pure analyte) is added in order to check analyte recovery. Again recoveries of between 98 and 102% would normally be expected. With spiked samples we require the additional analyte to be evenly distributed throughout the sample matrix. This can obviously cause problems with certain solid matrices, and thus the technique of spiking is more often employed with liquid or gaseous matrices.

11.3. STANDARD METHODS OF ANALYSIS

There are many books listing standard methods of analysis. All the more recent texts which incorporate 'standard' within the title are published by learned societies or government registered bodies. Examples:

- *Official Standardised and Recommended Methods of Analysis, The Royal Society of Chemistry, 1973*

- *British Pharmacopoeia* – The British Pharmacopoeia Commission;

- *Official Methods of Analysis of the Association of Official Analytical Chemists*, 14th Edition, 1984.

All these books list analytical methods resulting from collaborative work which has been carried out over a number of years by specialists in the topics concerned. All the methods quoted have been well tried and tested, and are capable of providing accurate results within the sample matrices for which they were developed.

We should emphasise, however, that the methods designated as 'Standard' or 'Official', represent only a 'state-of-the-art' situation, modification of existing procedures and addition of new procedures being an on-going process.

Much of the collabrative work in the U.K which leads to the designation of a method as 'standard' is instigated by the Royal Society of Chemistry through its Analytical Methods Committee. This committee, consisting of eminent analytical chemists from industrial, governmental and academic institutions, has two main responsibilities:

- to test and investigate new methods of analysis, with a view to designating the method as 'standard',
- to compare methods of analysis for use with difficult matrices.

The work is carried out by experienced analytical chemists working in laboratories which have proven expertise in that field. For instance, for testing of methods for the determination of traces of pesticides, at least one of the laboratories involved would be part of the ADAS group. [ADAS stands for Agricultural Development Advisory Service, which is part of the Ministry of Agriculture, Fisheries, and Foods.]

11.3.1. Requirements of a Standard Method of Analysis

Although it is difficult to define precisely what is meant by a standard method of analysis, the requirements of the method can be listed in order that it may be designated as standard.

- The method is known to be widely applicable to the analyte, present in a range of matrices.
- The method has been validated through inter-laboratory collaborative studies and is known to give accurate results.
- Interference effects are well documented.
- The method requires the use only of equipment which it would be usual to find within the average analytical laboratory.

11.4. ASSESSMENT OF ANALYTICAL PERFORMANCE

In considering both reference material and standard methods of analysis, we have introduced the terms *collaborative study* and *inter-laboratory study* as the most effective way of guaranteeing the performance of a method or the results of an analysis. Both these terms represent forms of comparative study.

There are three ways in which a comparative study may be carried out.

- between analysts, ie by intra-laboratory comparison,
- between laboratories, ie by inter-laboratory comparison,
- comparison between methods.

The comparison of different methods for carrying out an individual analysis can form part of either an intra- or an inter-laboratory comparative study.

Let us now consider these terms in a little more detail.

11.4.1. Intra-laboratory Comparison

Intra-laboratory testing provides a continuing assessment of the performance of both individual analysts and of laboratory instrumentation. Allowing two or more analysts to analyse the same sample by either the same or different procedures will help to identify any individual bias in the way in which a method is carried out, or to identify flaws in a given procedure.

SAQ 11.4a

Two analysts are asked to analyse the same sample by different procedures, each performing the analysis 10 times. If they obtain mean values which are significantly different from one another, but at the same time obtain results with equivalent precision, which of the following conclusions can we draw from their results?

⟶

SAQ 11.4a (cont.)

(*i*) One of the methods contains an indeterminate error.

(*ii*) Each method contains an indeterminate error.

(*iii*) One of the methods contains a determinate error.

(*iv*) Each method contains a determinate error.

(*v*) One of the results is inaccurate.

(*vi*) Both results are inaccurate.

When the results of an analysis are likely to form part of a litigation issue, then it is essential that the sample is analysed by two methods usually based upon different physical principles.

11.4.2. Inter-laboratory Comparison

Inter-laboratory testing, sometimes termed proficiency testing, is a programme of work whereby a single sample (or possibly a number of samples) is analysed in a number of different laboratories. The main objectives of the programme are:

- to provide a measure of the precision and accuracy of the analytical method as applied in different laboratories,
- to estimate the accuracy and precision of results as between laboratories,
- to identify weak methodology,
- to detect training needs,
- to upgrade, if necessary, the overall quality of laboratory performance.

The testing programme has a co-ordinating organisation/that provides the sample to be examined by the participating laboratories. These samples are generally homogeneous, in order to avoid any sampling errors, and the analytical procedure to be followed may also be specified.

Collaborative testing is a special form of inter-laboratory testing. This type of testing is used to evaluate an analytical method under real working conditions. All the participating laboratories follow the same procedure, and analyse portions of carefully prepared homogeneous samples.

List of Objectives

When you have completed this section you will,

- be aware of the importance of carrying out reference checks on analytical methods;

- be aware of the existence of reference and certified reference materials, and understand the difference between them;

- know of the existence of standard and official methods of analysis;

- appreciate the value of inter- and intra-laboratory comparisons.

Self Assessment Questions and Responses

SAQ 1.4a

Mark the following as either true or false.

(*i*) The highly sensitive methods available to modern analysts have minimised the problem of complicated sampling procedures.

True / False

(*ii*) Modern high-precision techniques can be applied only to homogeneous samples.

True / False

(*iii*) Large sampling errors are not removed by high-precision laboratory methods.

True / False

(*iv*) Sampling gradients should always be thoroughly investigated before laboratory determination.

True / False

⟶

SAQ 1.4a (cont.)

> **(*v*)** **Weighing is always preferred to volume measurement because it is inherently more precise.**
>
> **True / False**
>
> **(*vi*)** **Wet samples should always be rendered anhydrous before measurement or determination.**
>
> **True / False**

Response

Your response should look like this

		True	False
(*i*)	....		1
(*ii*)	....		1
(*iii*)	....	1	
(*iv*)	....		1
(*v*)	....		1
(*vi*)	....		1

Comments on responses

(*i*) The highly sensitive methods available in many modern analytical methods have in fact high-lighted the potential for changes in sampling to influence the results, hence the statement is false. What highly sensitive methods have done in terms of minimising complicated procedures is to eliminate some of the preconcentration steps which less sensitive methods need.

(*ii*) There are times when heterogeneous samples must be analysed, in which case precise methods are better than imprecise methods. A sufficient number of determinations coupled with statistical treatment should overcome the problem.

(*iii*) This is a major truism of chemical analysis. There is no way around the problem of large sampling errors other than the collection of a large number of samples (note: – not doing a large number of determinations on the same sample), and hoping that the errors are uniformly scattered around the true value.

(*iv*) There are very many occasions in industry and in public utilities when the accuracy required of a particular analysis is low and small sampling gradients would be well within acceptable limits. The investigation of gradients in sampling will generally require methods capable of much greater reliability than the measurement of any one sample alone, consequently such investigations will add cost to the analysis.

(*v*) Here, as in response (*iv*), the statement is false because of the word 'always'. If high precision is not required the convenience of volumetric methods often makes them more desirable.

(*vi*) The presence of water in samples may be of little importance, or it may be of vital importance to the user of the analytical results. The sort of bland oversimplification in (*vi*) is a result of failing to think of the context of the analysis, (*vi*) is false.

****

SAQ 2.a

> **Can you list some important factors that may need to be considered when deciding upon the sample size?**

Response

The list below includes most of the parameters that you will need to take into account when deciding upon the size of sample to be taken.

— The mass of material available for testing.
— The state and composition of the sample (ie dimensions, shape, uniformity of distribution, and specific mass of the granules).
— The proportion of the component to be determined.
— The sensitivity of the analytical method to be used.
— The accuracy demanded of the answer.
— The cost of the sample

If you managed to think of some of these then you are beginning to appreciate some of the problems associated with sampling.

****

SAQ 2b

> **Why do you consider handling and/or storage of a sample may cause problems?**

Response

If we return to the definition of a representative sample: a portion of a material taken from a consignment and selected in such a way that it possesses the essential characteristics of the bulk; then you will realise that any changes that occur to the sample during handling or storage will result in the sample analysed no longer being representative of the material from which it originated. Changes that may occur include:

— loss of fine particles,
— loss of moisture or solvent,
— loss of components due to adsorption on the wall of the container,
— changes due to exposure to air or light,
— absorption of moisture.

Therefore, when a sample is taken, you must consider the nature of the vessel in which it is to be stored, bearing in mind, before the final choice is made, all the factors listed above.

****

SAQ 3.1a

Five samples are taken from a single lot of material, and on analysis are found to contain on average 5.20% of the component of interest. In previous experiments where lots of similar material have been sampled thirty or more times, the measured sampling variance has been calculated to be 0.16. Which of the following results should the analyst report for this analysis, given that he wishes to be 95% confident that his reported results are correct:

(*i*) between 4.70% and 5.70%,
(*ii*) between 5.06% and 5.34%,
(*iii*) between 4.85% and 5.55%?

Response

We have been told in the question that five samples were taken and analysed giving a mean value for the component of interest of 5.20%. We must assume that the distribution of the component of interest throughout the lot sampled is similar to that found when similar lots have been analysed. We can use therefore the sampling variance established during the earlier experiments.

By using Eq. 3.5 we can calculate the value of E at the 95% confidence level, and this represents the spread of values around the calculated mean, within which we are 95% confident that our answer is correct.

$$n = [tS/E]^2 \tag{3.5}$$

$$= t^2S^2/E^2$$

where t, S, E relate to the terms given in Section 3.1 of this Unit.

$$\therefore \quad E^2 = [t^2S^2/n]$$

or by replacing S^2 by variance V

$$E^2 = [t^2V/n]$$

From the t-test table supplied we find that for 30 degrees of freedom, the t value at the 95% confidence level is 1.96.

$$E^2 = [1.96^2 \times 0.16/5]$$

$$E = 0.35$$

Thus the result that the analyst should quote will be 5.20 ± 0.35% ie between 4.85% and 5.55%

Therefore (*iii*) is the correct answer.

If the given sampling variance of 0.16 had related to the five samples analysed, rather than the extensive analysis carried out previously, then the value of E would have been,

$$E = [2.78^2 \times 0.16/5]^{\frac{1}{2}}$$

$$= 0.50$$

resulting in a spread of results between 4.70% and 5.70% [Answer (*i*)].

Remember, that when using the t test table for a given number of degrees of freedom <15, the number of degrees of freedom is reduced by one when the standard deviation is based upon an estimated mean rather than a true mean. Therefore, the value of 2.78 used in the calculation above is the t value for 4 degrees of freedom at the 95% confidence level.

If you carried out the calculation by using Eq. 3.5 but mistook the sampling variance for standard deviation,

ie $$E = [1.96^2 \times 0.16^2/5]^{\frac{1}{2}}$$

$$= 0.14$$

you would have obtained a range of results between 5.06% and 5.34% [answer (*ii*)].

If you were able to answer the question correctly, then you are beginning to appreciate the application of statistics to sampling. If you were not able to obtain the correct answer then return to Section 3.1, and read through it once more.

****

SAQ 3.2a

The following terms relating to sampling have been used within this section:

- **analysis sample,**
- **gross sample,**
- **increment,**
- **lot,**
- **sampling unit,**
- **sub-sample,**
- **composite.**

Relate these to one another, and suggest at which sampling stage the maximum error is likely to occur.

Response

The batch of material to be sampled is known as the lot, and may be divided into one or more sampling units. Samples removed from some or all of these sampling units are known as increments, and maybe combined together to form a gross sample. The term used to denote the combined increments is 'composite sample'. If the gross sample is too large to send for analysis, it will be sub-sampled one or more times until a sample small enough to submit for analysis is obtained. The material submitted for analysis is also sampled to obtain an analysis sample.

When a sample is removed from either the gross sample or any subsequent sub-sample we take care to ensure that the sample is representative. It is, however, less easy to take the same care when obtaining the initial increment, a problem on which we shall enlarge in Section 3.3 of this Unit. The maximum error therefore is most likely to occur at this stage of the sampling.

****

SAQ 3.4a

Explain the relationship between Eq. 3.5

$$n = [tS/E]^2 \qquad (3.5)$$

and Eq. 3.12

$$n = 4V/A^2 \qquad (3.12)$$

where n **is the number of samples taken,**

t **is the probability factor obtained from the t-test table,**

E **is the maximum allowable error,**

V **is the sampling variance,**

A **is the required sampling accuracy.**

Response

Eq. 3.5 and Eq. 3.12 both relate the number of samples to be taken to other variables.

Let us start by rearranging Eq. 3.5 to give,

$$n = t^2 S^2 / E^2$$

Given that:

(*a*) two standard deviations on either side of the mean value embrace approximately 95% of the statistical sample,

(*b*) the standard deviation for sampling was calculated after a large number of samples had been taken, and the analysis carried out on each sample taken,

the value of t becomes 1.96, which is approximately 2

$$\therefore \qquad t^2 \simeq 4$$

The sampling variance V is equal to the (standard deviation)^2,

$$S^2 = V$$

The accuracy A has been defined in Section 3.1 as being equal to two standard deviations, which is also the value of E at the 95% confidence level.

Therefore $n = t^2 S^2 / E^2 = 4V/A^2$

Under the conditions detailed in the question, the two equations are equivalent.

****

SAQ 4.0a

Can you list some of the main considerations that should be borne in mind when taking a sample and submitting it for analysis?

Response

The main considerations that we need to bear in mind when sampling are as follows:

- who is to take the sample?
- is the sample to be taken from a hazardous location, and therefore, do we need to take any safety precautions?
- are the sampling instruments clean and appropriate for the task in hand?
- do we have a reliable sampling procedure?
- how are we going to store the sample, for submission for analysis?
- have we labelled the sample correctly, including all the necessary information?

Well done, if you managed to remember all of these important points.

****

SAQ 4.1a

In what ways may we take a sample from a conveyor belt?

Response

There are a number of ways to take a sample from a conveyor belt, depending on whether the belt is moving or stationary, and whether the sample is to be taken manually or automatically.

If the conveyor belt is stationary, then we can remove a sample by taking a cross-section of material from the belt. The width of the sample removed will be determined by the variance expected for the material being sampled, and will certainly not be less than two to three times the diameter of the largest particle.

If the conveyor belt is in motion, then we have two options open to us to obtain a sample:

— to remove a diagonal cross-section from the moving belt, usually by using an automatic device;

— to remove a sample from the end of the conveyor belt where there is a free-fall situation. This may be done manually by using a large scoop, or automatically by diverting the falling material into a separate container.

Again, it is important to stress that however the sample is taken, it should be a cross-section of the material travelling along the belt.

****

SAQ 4.1b

What sort of equipment or method would you use to take samples from the following:

(*i*) a dry sandy soil,
(*ii*) a compacted sewage sludge,
(*iii*) a 25 kg bag of smokeless fuel?

Response

(*i*) A dry sandy soil.

A dry sandy soil is probably best treated as a particulate material containing a variety of sizes of particles. Provided none of the particles is very large, then some form of thief device (eg of the concentric-tube type) might prove useful. If the thief device proves to be inadequate however, then it will be necessary to remove by digging, a fairly large quantity of the soil, to break down the larger particles, and to sub-sample it by using the coning and quartering method.

(*ii*) A compacted sewage sludge.

Assuming that the sewage is wet, then a bayonet type of sampler would prove to be a useful tool in this context. The sample removed should be transferred immediately to a sealed container and sent to the laboratory for analysis in an 'as sampled' condition.

(*iii*) A 25 kg bag of smokeless fuel.

The sample to be tested is a relatively small amount of a material consisting of a variety of particle sizes. Probably the most suitable procedure would be to empty all the material on to a clean flat surface, and to reduce the average particle size. The sample could now be sub-divided by riffling, provided that a suitable riffler is available. Once a small representative sample has been obtained from the initial 25 kg bag, it can be ground and sieved, ready for analysis.

Finally, may I remind you, that there is no strictly right or wrong way to obtain a sample, provided, of course, that you obey the golden rule, that the sample taken is representative of the lot from which it originated. Sampling procedures are based upon practical experience, and if your answer did not quite agree with mine then don't despair, it may be equally as good.

****

SAQ 4.2a

What precautions must we take when sampling liquids that are flowing slowly?

Response

Liquids which flow slowly are subject to laminar flow. This means that the speed of flow differs across the flowing body, the flow being fastest at the centre of the body and effectively zero at the edges. We must take this fact into account when we attempt to take a representative sample.

For liquids flowing in pipes, we can homogenise the liquid just before we sample it by creating turbulence in the flow. This we can achieve in a number of ways, two of which were described in the text.

For wider bodies (eg rivers and canals), where it is not feasible to create turbulence, it will be necessary for us to take samples from several points across the body of water. Portions of the samples taken, may then be combined, in the ratio of the rates of flow at the points where the samples were taken.

****

SAQ 4.2b

What differences in sampling procedure would you adopt if you were required to sample a liquid of low viscosity inside two drums, one of which was of 10 dm^3 capacity and the other of 200 dm^3 capacity.

Response

The significant differences between these two samples is the size of the container. The container of 10 dm^3 capacity is small enough to pick up and shake to mix the contents, whereas the other container is too large to be treated in this way. So, in order to sample from the smaller vessel, we need only to shake the vessel thoroughly and immediately pour some of the mixed liquid into a glass bottle or other glass container. After allowing the sample to settle, we can ascertain whether the liquid in the drum consisted of one or more phases. Assuming it to be a single-phase system, we can now store the sample for analysis.

With a 200 dm^3 capacity drum, we have no simple way of finding out whether the liquid inside the drum consists of more than one phase, or of homogenising it. We must therefore sample the whole depth of the liquid, which we can do by using the sampling device illustrated in Fig. 4.2d. The sample removed can then be transferred to a suitable glass bottle, for example, for storage. With the sampling device illustrated, we may not be able to take a sample from the base of the drum, and thus it would be worthwhile, simply as a precaution, also to take a sample from the base of the drum by using the device illustrated in Fig. 4.2c.

****

SAQ 4.3a

Why are gaseous samples, more likely to suffer a loss of sample integrity on storage, than either solid or liquid samples?

Response

There are three reasons for gaseous samples on storage being likely to suffer a loss of sample integrity. These are:

- preferential adsorption of gases onto the wall of the container,
- preferential losses of gases by diffusion through plastic bags,
- changes in the composition of the sample caused by reduction in temperature and/or pressure.

Provided the gases being stored are not acidic, then storage in metal containers is to be preferred. Acidic components in metal containers are likely to be lost by reaction with the metal and to cause corrosion if the gas sample is at all wet. Some other gases may also react with metal surfaces, which tend to be less inert than those of glass or plastic materials. Glass suffers mainly from its ability to adsorb some gases preferentially, whilst plastics are permeable to some gaseous materials.

As we reminded you in the text, gases can alter in their overall composition due to changes occurring in their temperature and/or pressure. If, for instance, the gaseous sample was taken from a process stream which was at a temperature considerably higher than ambient, then if we allow the sample to cool down to room temperature, condensation of the less volatile components (eg water), is likely. Therefore, in order that the sample may return to its original state prior to analysis, it will be necessary to heat the sample until equilibrium is achieved. To be able to do this we must take care to choose a suitable sample container. Gases which are present at high pressures when they are sampled, are best stored at the same pressures to alleviate similar problems of condensation.

****

SAQ 4.3b

In the text we have divided atmospheric sampling locations into three categories:

- **narrowly defined areas,**
- **large enclosed areas,**
- **open atmospheres.**

In what way are these three divisions different as locations to be sampled?

In answering this question you should take into account the representative character of the sample both with respect to cross-section and to time, and whether the atmosphere was static or moving.

Response

In narrowly defined atmospheres, (eg at the top of a chimney stack), the gases are moving quite rapidly and so a single sample will probably be representative with respect to cross-section, of the gases emerging from the stack. In large enclosed atmospheres, a single sample is likely not to be representative of the whole, unless the gases have been physically mixed. If the atmosphere is static, which is not true in a factory environment, then an equilibrium gas-concentration may certainly be achieved eventually by the process of natural diffusion, and a single sample would then be representative of the whole.

Open atmospheres are subject to variations in the weather (eg wind and rain), and therefore a single sample will relate only to that atmosphere prevailing at the sampling point, at the time the sample was taken.

****

SAQ 5.3a

Organo-sulphur compounds on combustion in oxygen, produce an equilibrium mixture of sulphur dioxide and sulphur trioxide according to the following equations:

$$\text{'S'} + O_2 \rightarrow SO_2$$

$$SO_2 + \tfrac{1}{2} O_2 \rightarrow SO_3$$

The position of equilibrium depends upon experimental conditions, but a mixture is produced containing approximately 80% of the available sulphur as sulphur dioxide. This sulphur dioxide can be oxidised with iodine according to the following equations. ⟶

SAQ 5.3a (cont.)

> $$SO_2 + H_2O + I_2 \rightarrow SO_3 + 2H^+ + 2I^-$$
>
> $$SO_3 + H_2O \rightarrow H_2SO_4$$
>
> **The reaction occurs rapidly and stoichiometrically.**
>
> **Do you consider this analytical method for the determination of the sulphur to be absolute or comparative?**

Response

The strict answer to this question, is that the method is a *comparative* one, although you may well have said absolute, if you were considering only the final iodine titration at the end. As the production of sulphur dioxide depends on experimental conditions, you will need to find the value of the equilibrium constant, and thereby the percentage of SO_2 in the mixture of SO_2 and SO_3 produced, for the given set of experimental conditions. Once this value has been calculated, you can then relate the quantity of sulphur found from the iodine titration to the amount present in the original sample by the percentage yield obtained from oxidation.

****

SAQ 5.3b

> **In a typical spectrophotometric determination, the relationship between measured absorbance and concentration of analyte being determined is given by:**
>
> **Absorbance = k (concentration)**
>
> **where k is the proportionality constant.** ⟶

SAQ 5.3b (cont.)

> **If a solution containing 0.0100 mol dm^{-3} of the analyte gave an absorbance reading of 0.238, what would you expect the absorbance reading to be for a solution containing 0.0180 mol dm^{-3} of the analyte?**
>
> **Also, if absorbance is dimensionless, what are the units in which k is measured?**

Response

Spectrophotometric techniques are typical examples of comparative methods of analysis. In this example you are told that the relationship between absorbance (A) and concentration (C) is linear.

ie $$A = kC$$

Therefore we can initially find k.

$$k = \frac{A}{C} = \frac{0.238}{0.0100} = 23.8$$

Note the units for k are the reciprocal of those for concentration, ie mol^{-1} dm^{3}.

Now we can use the calculated value of k to find absorbance when C = 0.018 mol dm^{-3}.

$$A = 23.8 \times 0.018 = 0.428$$

****

SAQ 5.3c

> **Calculate the weight of zinc required to prepare 250.0 cm^3 of a solution containing 1000 ppm of Zn^{2+}.**

Response

To prepare 250.0 cm^3 of a 1000 ppm Zn^{2+} solution you will require 0.2500 g of zinc metal.

You remember from the text, that one ppm can be defined as one mg dm^{-3}. Therefore to prepare 1.00 dm^3 of 1000 ppm solution you will require 1000 mg = 1.000 g. Hence to prepare 250.0 cm^3 you will require 1/4 of this, ie 0.2500 g.

****

SAQ 5.3d

The amount of an analyte present in a material is known to be 25.00%. Within what limits would you expect to obtain a result for this analyte if the analysis is carried out:

(*i*) by using a gravimetric method,

(*ii*) by using a suitable optical method?

Response

(*i*) If you assumed the relative accuracy obtainable with a gravimetric method to be ±0.1%, then your answer should have been: 24.98 to 25.02%

This is calculated as follows:

$$0.1\% \text{ of } 25.00 = \frac{0.1}{100} \times 25.00$$

$$= 0.025 \text{ (say 0.02)}$$

The answer therefore becomes 25.00 ± 0.02

$= 24.98$ to 25.02

If, on the other hand, you chose a relative accuracy of $\pm 0.2\%$, by a similar method your answer should now be 24.95 to 25.05%

(*ii*) For the optical method, the answer will be less accurate, as *all optical methods are comparative*, and thus cannot be expected under normal circumstances to achieve an accuracy better than $\pm 1\%$ of the true value. The expected answer is therefore 24.75 to 25.25%.

****

SAQ 5.4a

You are asked to prepare four litres of a standard solution of sulphuric acid (approximately 0.1 M) from an AnalaR sample of the concentrated acid. Which if any of the following procedures would be suitable?

[Concentrated sulphuric acid contains 96% w/v H_2SO_4; density conc. H_2SO_4 = 1.84 g cm^{-3}, $M_r(H_2SO_4)$ = 98.0].

(*i*) By using a measuring cylinder, transfer 22 to 23 cm^3 of the concentrated sulphuric acid into a large beaker containing about 2 litres of distilled water. Stir the solution continuously whilst adding the acid, and then allow the dilute acid to cool to room temperature before transferring it quantitatively to a 4 litre volumetric flask. Dilute the resultant solution to volume with distilled water. ⟶

SAQ 5.4a (cont.)

> **(*ii*) By using a calibrated bulb pipette, transfer carefully 22.0 cm^3 of the concentrated sulphuric acid into a large beaker containing about 2 litres of distilled water. *Remainder of procedure as per (i) above.***
>
> **(*iii*) Weigh out accurately 40.333 g of concentrated sulphuric acid into a 4 litre beaker containing about 2 litres of distilled water. *Remainder of procedure as per (i) above.***

Response

(*i*) This is by far the quickest method of preparing a standard solution of sulphuric acid.

Given that M_r (H_2SO_4) = 98.0, for a 0.1 M solution you will therefore require:

9.80 × 4 = 39.2 g for 4 litres of dilute solution.

However, the concentrated acid is only 96% pure.

∴ the weight of the given acid required is:

$$39.2 \times \frac{100}{96} = 40.83 \text{ g}$$

As the density of the acid is 1.84

$$40.83\text{g} = \frac{40.83}{1.84} \text{ cm}^3$$

$$= 22.2 \text{ cm}^3$$

A volume of between 22 to 23 cm^3 would therefore be needed.

Having prepared the dilute acid you may now like to think of a couple of suitable standards which could be used to measure accurately the molarity of your prepared solution. Check your answers against the list given in Fig. 5.4a of this Unit.

(*ii*) and (*iii*) Unfortunately concentrated sulphuric acid is *not* a primary standard, and therefore you cannot prepare a standard solution of the acid directly from the reagent either by weighing or pipetting it. The reasons for this are as follows.

– Concentrated sulphuric acid is hygroscopic (it is used as a desiccant!). The 96% figure quoted is therefore only approximate. You would also find difficulty in weighing accurately the amount quoted.

– Have you ever seen a 22.0 cm^3 total-volume pipette? I rather doubt it. Therefore, the acid would have to be delivered by a graduated pipette probably orginally calibrated to deliver water. Because of the viscosity of concentrated sulphuric acid is much greater than that of water, it would be virtually impossible to deliver 22.0 cm^3 by using this pipette. Of course, either method would give a solution which could later be standardised by use of a suitable primary standard.

****

SAQ 5.4b

Which of the following reagents could be used to standardise an approximately 0.1 molar solution of dilute hydrochloric acid?

(*i*) Anhydrous disodium oxalate,

(*ii*) A solution of sodium hydroxide prepared from sodium hydroxide pellets,

(*iii*) Anhydrous disodium carbonate.

Response

(*i*) This is indeed a primary standard, but only for redox titrations. Hydrochloric acid will certainly react with the sodium oxalate to produce oxalic acid,

$$Na_2C_2O_4 + 2\,HCl \rightarrow 2\,NaCl + H_2C_2O_4$$

but the titration end point will be difficult to detect because of the buffering action of the oxalic acid generated during the reaction.

(*ii*) Sodium hydroxide pellets not only absorb water from the air but also carbon dioxide, and are thus not acceptable for the preparation of a standard solution of sodium hydroxide.

(*iii*) Of the reagents suggested for standardisation, only anhydrous disodium carbonate is acceptable. This can be obtained, or better prepared, in a pure state. It is not significantly hygroscopic, and is stable in air.

If your answer to the question was (*iii*) then you were correct – well done.

****

SAQ 6.1a

Fig. 6.1d below shows a section of a linear calibration plot. Is the sensitivity of the method, to which this calibration relates approximately:

(*i*) 2.10 μA ppm^{-1},

(*ii*) 2.81 μA ppm^{-1},

(*iii*) 0.54 μA ppm^{-1},

(*iv*) 0.48 μA ppm^{-1}?

→

SAQ 6.1a (cont.)

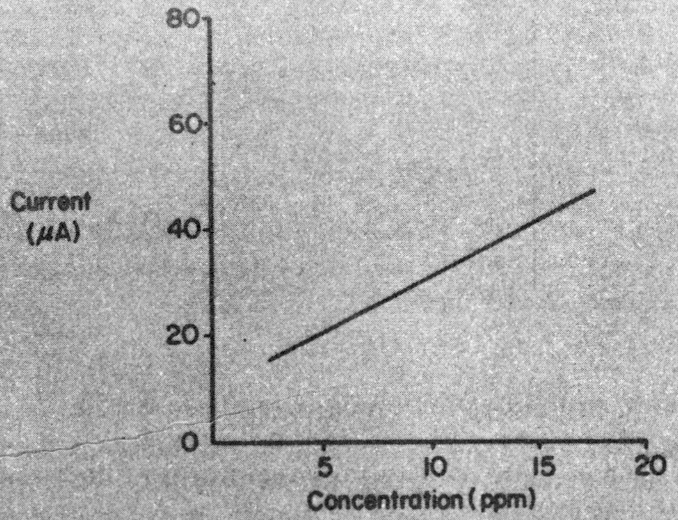

Fig. 6.1d. *Linear calibration plot for use in SAQ 6.1a*

Response

The sensitivity of an analytical method is normally quoted in terms of the change in the magnitude of the physical parameter per unit change in concentration. It can therefore be determined by measuring the *slope* of the calibration graph.

You will recall from the text however that the slope must be measured taking into account the units of the axes of the calibration graph. The slope can thus be obtained by choosing two suitable values of current or concentration and relating the concentration or current respectively to the chosen values.

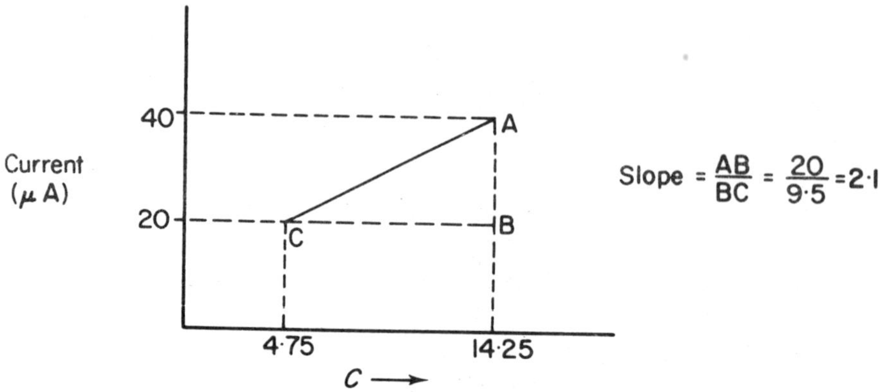

If you measured your slope as CB/AB, ie 9.5/20,

you have actually obtained the reciprocal of the slope,

ie 0.48.

If you forgot to take into account the units of the axes and measured the slope in purely arithmetic terms,

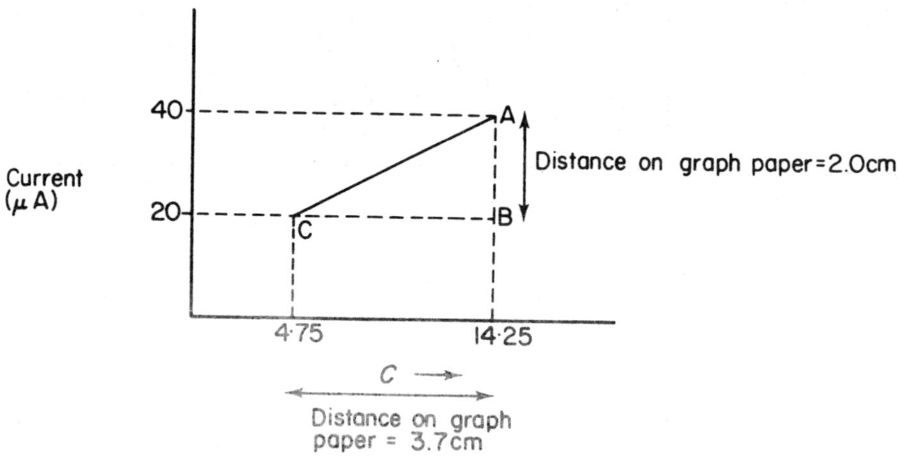

AB/CB now becomes 2.0/3.7 = 0.54.

If the straight line had passed through the origin, which, in the example quoted you can see does not occur, it would have been possible simply to choose a current value on the graph and divide it by its corresponding concentration. However where an analytical method has a determinate error, as in the example quoted, different answers will be obtained depending upon the values chosen for the calculation.

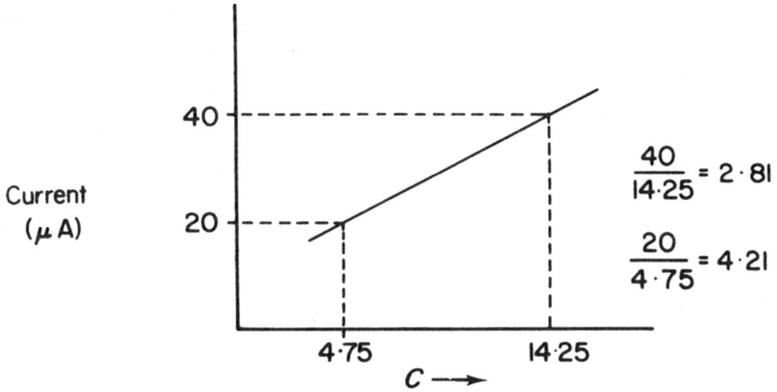

The answer to the question is therefore (*i*)

****

SAQ 6.2a

An analytical method is known to obey the relationship:

$$Y = be^{-mC}$$

(*A*) Does this method exhibit a relationship whereby either

(*i*) *Y* increases with increasing concentration, ⟶

SAQ 6.2a (cont.)

(*ii*) Y decreases with increasing concentration

Attempt to answer part (*A*) of this SAQ before continuing to part (*B*).

(*B*) The following set of data was obtained from an analytical method thought to obey the experimental relationship quoted above.

Y	C
2.26	2.00
1.97	4.00
1.67	6.00
1.45	8.00
1.23	10.00

What are your conclusions regarding this hypothesis?

Is the hypothesis

(*i*) definitely true,

(*ii*) probably true,

(*iii*) definitely false,

(*iv*) probably false?

Response

(*A*) The equation $Y = be^{-mC}$

is a typical equation for an exponential relationship and because of the negative sign in the index it tells us that there is an inverse relationship between Y and C.

Therefore Y decreases with increasing concentration

(B) The equation $Y = be^{-mC}$

can be converted into a linear relationship by taking logs of both sides of the equation.

Thus $Y = be^{-mC}$ becomes

$\ln Y = \ln b - mC$ (where ln = natural logarithm)

Now $\ln x = 2.303 \log x$

Therefore $2.303 \log Y = 2.303 \log b - mC$

or $\log Y = \log b - mC/2.303$

Thus there is a linear relationship between $\log Y$ and C with a slope of $-m/2.303$. Thus in order to prove or disprove the hypothesis that the data obey the exponential equation, we need first to take the logarithm of Y and plot this against C.

Y	$\log Y$	C
2.26	0.354	2.00
1.97	0.295	4.00
1.67	0.223	6.00
1.45	0.161	8.00
1.23	0.090	10.00

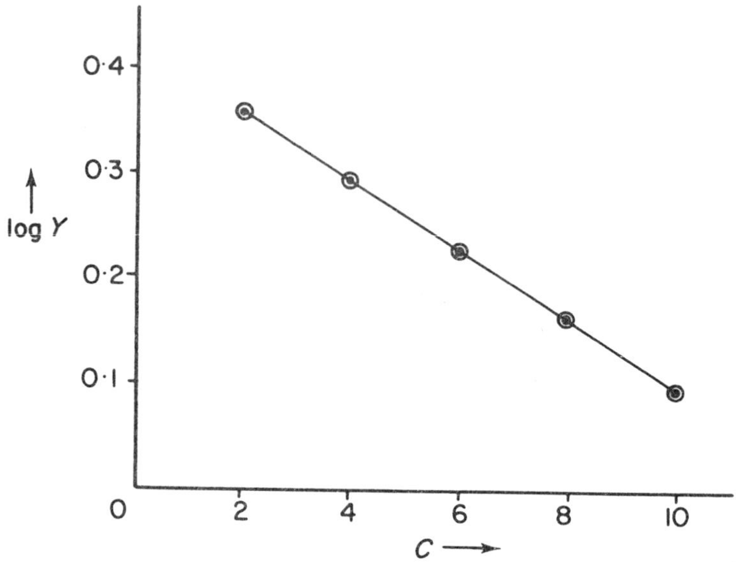

As we can see from the graph the data-points appear to fall on a straight line thus leading to the conclusion that the hypothesis *is probably true*.

We could alternatively have plotted a graph of ln Y against concentration which would have given us a similar result.

****

SAQ 6.4a

Which of the following properties are important in choosing a suitable standard for calibration in a comparative analytical technique?

(*i*) The standard substance should not react with any components normally present in air. ⟶

SAQ 6.4a (cont.)

> **(*ii*) Impurities present in the standard substance should not normally exceed 0.1%.**
>
> **(*iii*) The standard substance should be of a high relative molar mass in order to minimise weighing errors.**
>
> **(*iv*) The standard substance should be readily soluble in the solvent under the conditions in which it is to be used.**
>
> **(*v*) The standard substance does not necessarily have to be of an accurately known formula, so long as the percentage of the analyte constituent is accurately known.**
>
> **(*vi*) The standard substance should be present in an anhydrous state.**

Response

As was indicated in the text, the requirements for a standard which is to be used for a comparative analytical method are less stringent than those required by an absolute method.

(*i*) This is an important property. The standard should not be hygroscopic or react with carbon dioxide or oxygen. Whenever we have no choice in the standard substance and we are forced to choose one that does react with water, carbon dioxide or oxygen, then we should take precautions to remove these substances from the space where initial weighings are to be carried out.

(*ii*) As comparative methods are assumed accurate to only about ±1%, then a standard substance containing about 0.1% of impurities would be quite acceptable as a standard. This is therefore not an important property.

(*iii*) With comparative analytical methods, calibrations are normally achieved by diluting an initially prepared standard solution rather than by weighing out individual standards for each calibration point. Therefore we no longer need to restrict ourselves to standards of high relative molar mass to minimise weighing errors. Also, for organic analytes, the organic substance itself is the standard, and then we have no choice as to its relative molar mass. This is therefore an unimportant property.

(*iv*) Although this property is important it cannot always be realised. For instance, the most suitable standard for arsenic is arsenic (III) oxide (As_2O_3). This is insoluble in water, but can be dissolved slowly by boiling it with dilute aqueous sodium hydroxide, when it is converted into sodium arsenite. (Arsenic in solution present in its 3-valent state). From then on the resultant solution can be used as a standard for most determinations of arsenic in aqueous solution.

A problem that sometimes arises is the choice of a standard for use in the determination of trace metals present in organic media (eg traces of wear metals in engine oils). We have to use standards which are themselves soluble in the medium for which the analysis is to be carried out. This obviously excludes many compounds which normally form useful standards in aqueous solutions.

(*v*) There is no need for a standard to be of a known formula so long as we have an exact percentage of the component of that standard that is to act as the analyte. Thus, for metal standards for use in organic media, salts of the metal with complex organic acids can be purchased. These standards have a guaranteed percentage metal content, generally to 2 places of decimals, but may not be of accurately known formula.

(*vi*) There is no necessity to restrict ourselves to the use of anhydrous salts. Many hydrates are of accurately known formula and can be obtained highly pure.

Well done if you managed to get all of these parts correct.

****

SAQ 6.5a

A colorimetric method is to be used for the determination of trace quantities of palladium present in aqueous solutions. You have followed the recommended procedure, which involved the preparation of separate standard solutions and have thereby obtained the calibration data below.

Mass of palladium (W) in 500 cm^3 of measured solution (μg)	Absorbance (A)
5.0	0.225
10.0	0.315
15.0	0.375
20.0	0.470
25.0	0.550
Blank (0)	0.075

Attempt to answer the following questions related to the analytical method described above.

(*i*) Assuming that a linear relationship exists between absorbance and weight of palladium in solution, construct the best straight line through the data-points given. What is the exact equation for the relationship?

(*ii*) Is the analytical method exhibiting:
- a determinate error,
- an indeterminate error,
- both a determinate and an indeterminate error,

(*iii*) A sample containing an unknown quantity of palladium is analysed by the analytical method, giving the following results. ⟶

SAQ 6.5a (cont.)

Sample absorbance = 0.615

Blank absorbance = 0.070

What is the weight of palladium in μg in the sample analysed? Comment upon the validity of the results obtained.

(*iv*) Do you consider that the calibration could alternatively have been obtained by using an '*in-situ*' method?

(*v*) How many of the three abridged methods of calibration could be used with this analytical procedure? Suggest which they are and give reasons for your choice.

Response

As you are told that all weights of palladium relate to the same volume of solution, you can plot a graph directly of absorbance *vs* mass of palladium. However as the blank value may not be constant between batches of reagent, probably it is best to subtract the blank value of absorbance from the measured absorbances before plotting the results.

Mass (W) of palladium (μg)	Absorbance – Blank (A')
5.0	0.150
10.0	0.240
15.0	0.300
20.0	0.395
25.0	0.475

(*i*) In order to construct the best straight line through the data-points given, you will need to calculate the values of *m* and *b* in the equation for a straight line, viz

$$Y = mC + b$$

by the method of least squares, where

$$m = \frac{\Sigma C \Sigma Y - n\Sigma CY}{(\Sigma C)^2 - n\Sigma C^2}$$

$$b = \frac{\Sigma C \Sigma Y - \Sigma C^2 \Sigma Y}{(\Sigma C)^2 - n\Sigma C^2}$$

You will first have to calculate the various factors that are required in these equations.

Y	C	C^2	CY
0.150	5.0	25	0.75
0.240	10.0	100	2.40
0.300	15.0	225	4.50
0.395	20.0	400	7.90
0.475	25.0	625	11.88
$\Sigma Y = 1.56$	$\Sigma C = 75.0$	$\Sigma C^2 = 1375$	$\Sigma CY = 27.43$

$\therefore \quad (\Sigma C)^2 = 5625$

$n = 5$

$$\therefore \qquad m = \frac{(75 \times 1.56) - (5 \times 27.43)}{5625 - (5 \times 1375)}$$

$$= \frac{117 - 137.15}{5625 - 6875}$$

$$= \frac{-20.15}{-1250}$$

$$\therefore \qquad m = 0.016$$

$$b = \frac{(75 \times 27.43) - (1375 \times 1.56)}{-1250}$$

The denominator of b is the same as that for m.

$$\therefore \qquad b = \frac{2057 - 2145}{-1250}$$

$$= \frac{-88}{-1250}$$

$$= 0.070$$

The intercept for the best straight line is therefore 0.070.

You can now choose any mass value between 5 and 25 μg and by using the values of m and b already calculated, find the corresponding value of Y (Absorbance – blank) in this case.

It is preferable to choose large values in order to draw more accurately the best straight line.

Therefore let mass be 25 μg.

Now $Y = mC + b$

$$\therefore \quad A' = 0.016 \times 25 + 0.07$$

$$= 0.47$$

The straight line should therefore be constructed between

$W = 0, \; A' = 0.070$ and

$W = 25, \; A' = 0.47.$

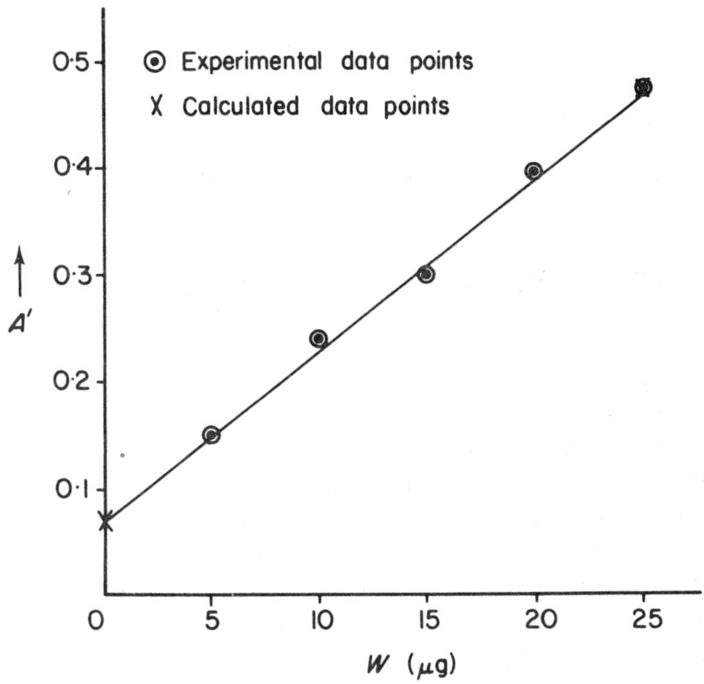

The exact equation for the relationship is therefore given by:

$$A' = 0.016W + 0.070$$

(*ii*) The method exhibits both a determinate and indeterminate error. It is most unusual for any comparative method to be totally free from indeterminate errors. Here it is obvious that there is an indeterminate error, in that it is impossible to construct a straight line through the data as orginally measured. Also because b has a finite value, after the blank value has been subtracted, a determinate error is indicated.

(*iii*) The answer is best obtained by using the equation relating to this analytical method,

ie $$A' = 0.016W + 0.070$$

As the sample absorbance = 0.615
and the blank absorbance = 0.070

$$A' = 0.545$$

∴ inserting this value into the equation we get

$$0.545 = 0.016W + 0.070$$

$$\therefore \quad \text{mass of palladium } (W) = \frac{0.545 - 0.070}{0.016}$$

$$= 29.7\ \mu\text{g},$$

The main feature that limits the validity of the result obtained is that *the calibration has been used beyond the calibrated region*. You have no knowledge as to the extent of linearity for this method, and therefore the sample should have been diluted before analysis to bring the result within the calibrated region of 5 to 25 μg. If anything, the result may be a little low because of a slight negative deviation.

(*iv*) The technique of '*in-situ*' calibration is not used normally for spectrophotometric analysis, although if performed carefully, so as to avoid any losses of solution, could certainly be used in this context. Supposing you had prepared the initial reagent solution and diluted this to 50.0 cm^3. As the volume of a cell for spectrophotometric measurement is generally not more than a few cm^3, only a portion of the prepared solution would have been transferred initally to the cell to obtain the blank absorbance value. This would then need to be returned to the remaining solution before beginning the addition of the palladium standard. This process would have to be repeated for each standard prepared. To avoid any losses of solution, care would have to be exercised at each transference stage. Also the standard palladium solution would have to be added in microlitre quantities so as not to dilute the total solution significantly. The resultant calibration data would be less precise than those obtained by using the separate solution method.

(*v*) You will recollect from Section 6.3 that there are three methods of carrying out abridged calibration procedures:

— use of a single standard,

— use of a single standard and a blank,

— use of two standards and a blank.

As the method has been shown to have a significant blank value, together with a determinate error, only *one* of them would be acceptable. This is the third one, ie that involving the use of *two standards and a blank*.

****

SAQ 7.1a

A single standard addition procedure is to be used in an analysis involving a logarithmic relationship. If the two equations, before and after standard addition, are:

before addition, $Y_o = b - m \log C$

after addition, $Y_i = b - m \log (C + C_s)$,

(*i*) develop an equation which relates C to the other parameters in the equations;

(*ii*) having developed your equation, what extra information do you require in order to calculate C?

You may assume there are no significant dilution effects on addition of the standard.

Response

(*i*) In order to develop our equation we must solve the two simultaneous equations given below.

Since $$Y_o = b - m \log C \tag{1}$$

and $$Y_i = b - m \log (C + C_s) \tag{2}$$

We can eliminate b by subtracting Eq. (2) from Eq. (1)

Hence $$Y_o - Y_i = -m \log C - [-m \log (C + C_s)] \tag{3}$$

$$= m [\log (C + C_s) - \log C] \tag{4}$$

Now $$\log (C + C_s) - \log C = \log \frac{(C + C_s)}{C}$$

Thus Eq. (4) becomes

$$Y_o - Y_i = m \log \frac{(C + C_s)}{C} \tag{5}$$

ie $$\frac{Y_o - Y_i}{m} = \log \frac{(C + C_s)}{C} \tag{6}$$

Hence, $$\text{Antilog} \frac{(Y_o - Y_i)}{m} = \frac{(C + C_s)}{C} \tag{7}$$

Multiplying both sides of Eq. (7) by C, we get

$$C \text{ Antilog} \frac{(Y_o - Y_i)}{m} = C + C_s \tag{8}$$

Subtracting C from both sides gives Eq. (9)

$$C \text{ Antilog} \frac{(Y_o - Y_i)}{m} - C = C_s \tag{9}$$

which can be rearranged to give Eq. (10)

$$\text{Antilog} \frac{(Y_o - Y_i)}{m} - 1 = \frac{C_s}{C} \tag{10}$$

Dividing both sides of Eq. (10) by

$$\text{Antilog} \frac{(Y_o - Y_i)}{m} - 1$$

gives

$$C = \frac{C_s}{\text{Antilog} \frac{(Y_o - Y_i)}{m} - 1} \tag{11}$$

You could alternatively have subtracted Eq. (1) from Eq. (2), in which case you would have obtained the final Eq. (12).

$$C = \frac{\dfrac{\text{Antilog}\,(Y_i - Y_o)C_s}{m}}{1 - \text{Antilog} \dfrac{(Y_i - Y_o)}{m}} \tag{12}$$

Remember that the antilog of both

$$\frac{(Y_o - Y_i)}{m} \quad \text{and} \quad \frac{(Y_i - Y_o)}{m}$$

are pure numbers and that these values are easily obtained with modern scientific calculators. The value of C will actually be the same whichever route you use to obtain the equation.

If you did not manage to obtain the correct answer first time, then look very carefully at your signs. Remember for instance, that the product of two negative quantities is positive.

The equations you have just developed will be considered again in a later unit on potentiometry as they relate to the use of a selective-ion electrode, selective to a negative ion.

Also if you would like to prove that Eqs. (11) and (12) are in fact the same, use the following set of data.

Let $Y_o = 256$, $Y_i = 195$, $m = 59$ and $C_s = 0.10$.

The answer should be $C = 0.010$ in both cases.

(*ii*) Equations (1) and (2) possess three unknown variables, C, b, and m. By eliminating the constant term b as shown, we can obtain only equations which relate C to m, the slope of the linear equation. Therefore in order to perform this calculation, we need to know the value of m.

****

SAQ 7.1b

In the technique of anodic stripping voltammetry, the peak height of the anodic wave is proportional to the concentration of analyte according to the relationship:

$$i_p \propto C$$

where i_p refers to peak height and providing that all experimental conditions are held constant.

The following data relate to the determination of traces of lead in drinking water.

Volume of water sample analysed	**= 10.0 cm^3**
Initial peak height of the lead wave	**= 1.56 μA (i_{po})**
Volume of standard added	**= 1.00 cm^3**

→

SAQ 7.1b (cont.)

Concentration of lead in the standard	**= 0.100 mg dm^{-3}**
Peak height of the lead wave after addition of the standard	**= 3.79 μA (i_{pi})**

(*i*) Is the increase in concentration of the lead added to the analysis cell:

100.0 ppb,
or 10.0 ppb,
or 9.1 ppb?

(*ii*) Calculate the concentration of lead in ppb in the sample of drinking water.

Response

(*i*)

100.0 ppb is incorrect. The concentration of the lead solution from which the standard addition was made was 0.100 mg dm^{-3}. This is equivalent to 0.1 ppm or 100 ppb. However this undergoes a dilution of 1.0 cm^{3}, to 11.0 cm^{3}, which has not been taken into account in this answer.

10.0 ppb is incorrect. You have correctly calculated that 0.100 mg dm^{-3} equivalent to 100 ppb, and then remembered to account for the dilution effect. However the dilution was not 1 to 10 but 1 to (10 + 1).

9.1 ppb. If you picked this answer first time then you are a good way to understanding the standard addition procedure. The initial concentration of the lead standard was 0.100 mg dm^{-3} = 100 ppb, and 1.0 cm^{3} of this solution was diluted to (10 + 1) = 11.0 cm^{3}.

$\therefore$ The increase in concentration of lead caused by the addition of the standard was

$$\frac{1}{11.0} \times 100 \text{ ppb} = 9.09 \quad \text{ie } 9.1 \text{ ppb.}$$

(*ii*) Because of the similarity of all the standard addition equations, you are well advised to attempt the calculation from first principles, rather than attempting to remember the final equation.

In the example quoted there is a significant dilution of the analyte after addition of the lead standard. The two equations we require therefore are,

$$i_{po} = kC \tag{1}$$

and

$$i_{pi} = k\left\{\frac{C(V)}{V + v} + C_s\right\} \tag{2}$$

where k is the proportionality constant.

Substituting for k in Eq. (2) gives

$$i_{pi} = \frac{i_{po}}{C}\left\{\frac{C(V)}{V + v} + C_s\right\} \tag{3}$$

We can now insert some values into this equation

$$3.79 = \frac{1.56}{C}\left\{C\frac{10}{11} + 9.1\right\}$$

$$= \frac{1.56 \times 10}{11} + \frac{14.2}{C}$$

$$= 1.42 + \frac{14.2}{C}$$

$$\therefore \qquad C = \frac{14.2}{3.79 - 1.42}$$

$$= \frac{14.2}{2.37}$$

$$= 6.0 \text{ ppb}$$

****

SAQ 7.2a

In the polarographic determination of Cd^{2+} in an aqueous solution the following procedure was adopted. 25.0 cm^3 of a sample solution containing an unknown quantity of Cd^{2+} was pipetted into a polarographic cell and the diffusion current (i_d) was measured. To this solution was then added 1.0 cm^3 aliquots of a standard Cd^{2+} solution containing 100 ppm of Cd^{2+}, the diffusion current being measured after each addition of the standard solution. The results obtained are given in the table below.

Diffusion current, i_d (μA)	Volume of standard added (cm^3)
10.6	0
17.1	1.0
23.1	2.0
23.7	3.0
34.0	4.0

Assuming that all measurements have already taken into account any blank value, calculate the concentration of Cd^{2+} in the sample solution.

Response

We are told that the standard solution contains 100 ppm Cd^{2+} (100 ppm, remember, means 100 $\mu g\ cm^{-3}$). Therefore, by using the procedure outlined we can obtain the concentration of Cd^{2+} present in the unknown by plotting the corrected diffusion current against the weight of standard added.

Wt of std added (μg)	Diffusion current (μA)	Corrected diffusion current (μA)
0	10.6	10.6
100	17.1	17.1 × 26/25 = 17.8
200	23.1	23.1 × 27/25 = 24.9
300	28.7	28.7 × 28/25 = 32.1
400	34.0	34.0 × 29/25 = 39.4

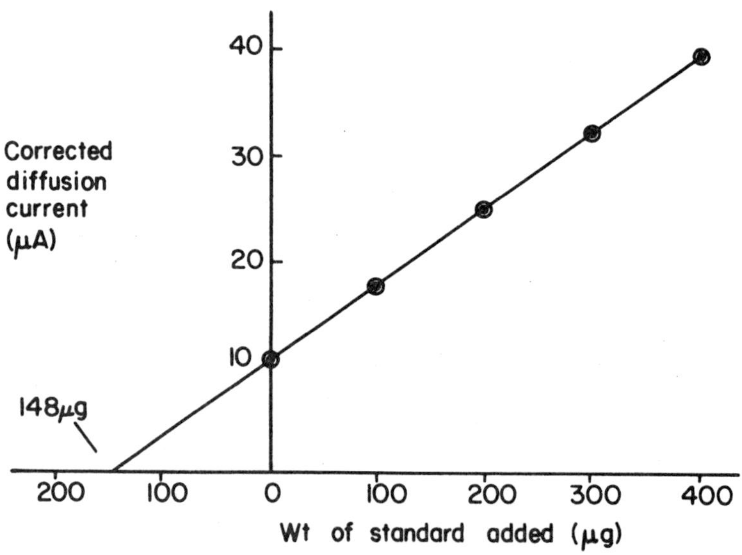

From the graph we see that the unknown sample contains approximately 148 μg in a total volume of 25.0 cm^3 of sample solution.

Thus the concentration of Cd^{2+} in ppm is

$$\frac{148}{25.0} = 5.9 \text{ ppm.}$$

****

SAQ 7.2b

The three graphs (A), (B) and (C), shown below, all relate to the same analysis performed by using the multiple standard-addition procedure. Can you identify, if and when an error has occurred in the analysis, and suggest possible sources of that error.

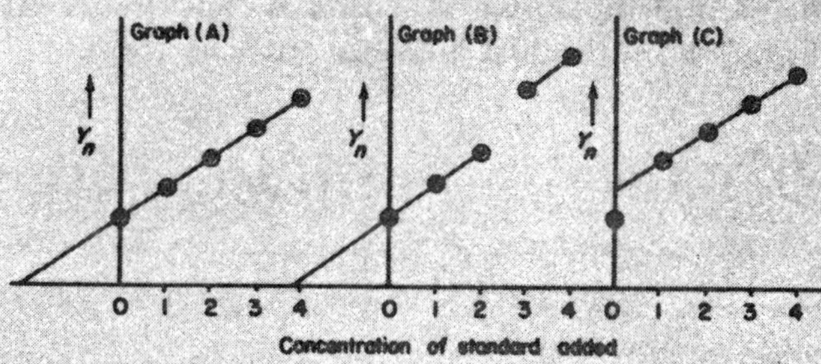

Response

Graph (A) This graph shows that the procedure appears to have been carried out correctly, and that the method is working.

Graph (B) Y_o, Y_1, and Y_2 data-points all appear to lie on a good straight line. Y_3 and Y_4 however appear to lie on a line parallel to the first but not coincident with it. This indicates that an error has occurred probably during the addition of the 3rd standard. The operator appears to have added too much standard at this point. The whole analysis should be repeated.

Graph (C) The points Y_1, Y_2, Y_3 and Y_4 appear to lie on a good straight line which does not include Y_o. The error in this case could be caused by one of two factors:

(*i*) An operator error during the addition of the first standard, too much standard being added.

(*ii*) A more fundemental error, in that the signal Y_O given by the measuring device was not due to the presence of the analyte, but was caused by the presence of some other interfering species masquerading as the analyte.

The interfering species is unlikely to exhibit the same sensitivity as the analyte (ie different slopes in the straight-line equation) and thus the point Y_o does not conform with the remainder of the data. The analysis should be repeated carefully, and if the effect is repeated, an alternative method of analysis must be used.

****

SAQ 7.3a

A sample of water known to contain sodium and potassium salts is to be analysed for its sodium content by atomic absorption spectroscopy. When this technique is used, potassium is known to interfere with the analysis by enhancing the sodium absorbance. If the sample is thought to contain between 100 and 150 ppm of sodium, suggest an optimum procedure for the determination by using a single standard-addition technique.

The calibration data obtained from a previous experiment for the determination of sodium in the presence of potassium are given in the table below. ⟶

SAQ 7.3a (cont.)

Concentration of sodium (ppm)	Absorbance (A)
1.0	0.170
2.0	0.335
3.0	0.505
4.0	0.640
5.0	0.750

Response

We first need to plot the calibration data given in order to determine the extent of linearity for this analysis.

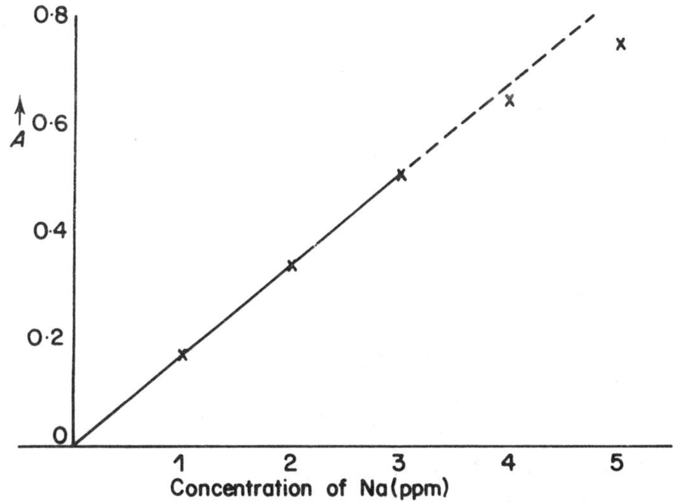

As we can see, the method appears to show linearity up to 3 ppm of sodium. The maximum concentration of sodium in the sample plus standard solution should therefore not exceed this figure.

Also bearing in mind that we should like the standard to increase the sample concentration by about 100%, we should aim to use a standard concentration of 1.5 ppm. The sample is thought to contain between 100 and 150 ppm of sodium, and thus effectively requires diluting 100-fold for analysis. This should preferably be carried out by successive dilution, ie dilute 10.0 cm^3 of the water sample to 100.0 cm^3 (solution A). This solution now contains between 10 and 15 ppm of sodium. We now need to pipette 10.0 cm^3 of solution A into each of two 100 cm^3 volumetric flasks, and to one of the flasks add 150 μg of sodium as a salt. We can now dilute both to volume with pure water. In analysing both of the prepared solutions by atomic absorption spectroscopy we should be looking for a maximum absorbance not exceeding about 0.5.

The optimum procedure you should have suggested therefore should have contained the following main points.

- The water sample must be diluted about 100-fold before analysis.
- The standard addition concentration should be 1 to 1.5 ppm.
- The absorbance of the analysed solutions should not exceed about 0.5.

If you managed to get all of these correct, then you have understood the fundamental principles of standard addition.

****

SAQ 8.1a

In the determination by hplc of the vitamin C and the saccharin content of a soft-drink concentrate, caffeine was used as the internal standard. The procedure used is outlined below.

A standard solution was prepared which contained the following concentrations of the three compounds: ⟶

SAQ 8.1a (cont.)

Vitamin C (250 ppm w/v), saccharin (100 ppm w/v), and caffeine (150 ppm w/v).

A portion of the standard solution was analysed by hplc and the resultant chromatogram was displayed by means of a potentiometric chart recorder. The peak heights for the three compounds were measured and the results obtained are listed below.

Compound	Peak height (cm)
Vitamin C	15.15
Saccharin	10.50
Caffeine	12.25

20.0 cm^3 of a soft-drink concentrate was pipetted into a 100.0 cm^3 volumetric flask, and to it was added 5.00 mg of pure caffeine. The mixture in the flask was then diluted to volume with distilled water. On analysis by hplc under conditions similar to these used for the initial standardisation, the following peak heights were obtained for the three compounds of interest.

Compound	Peak height (cm)
Vitamin C	6.95
Saccharin	5.00
Caffeine	8.30

Assuming that over the concentration range examined, a linear relationship exists between the measured peak height and concentration for all three compounds, calculate the concentration ($mg\ dm^{-3}$) of vitamin C and of saccharin in the soft-drink concentrate.

Response

The analysis described in this SAQ is to be carried out by using a single internal-standard method. We therefore need to calculate the response factors for vitamin C and caffeine, and for saccharin and caffeine. These can be obtained from the first set of analytical data.

$$\text{Peak height for vit C} = R\, C_{\text{vit C}}$$

$$\text{Peak height for caffeine} = R'\, C_{\text{caf}}$$

$$\therefore \quad \frac{15.15}{12.25} = \frac{R\,250}{R'\,150}$$

$$\therefore \quad \frac{R}{R'} = \frac{15.15 \times 150}{12.25 \times 250}$$

$$\text{ie} \quad R_{\text{vit C/caf}} = 0.74$$

Similarly

$$\frac{\text{Peak height for saccharin}}{\text{Peak height for caffeine}} = \frac{R''}{R'}\frac{C_{\text{sac}}}{C_{\text{caf}}}$$

$$\therefore \quad \frac{10.50}{12.25} = \frac{R''}{R'}\frac{100}{150}$$

$$\therefore \quad \frac{R''}{R'} = \frac{10.50 \times 150}{12.25 \times 100}$$

$$\text{ie} \quad R_{\text{sac/caf}} = 1.29$$

We can now use these response factors to determine the concentration of vitamin C and of saccharin in the soft-drink concentrate.

(*i*) Vitamin C

The concentration of vitamin C present in the sample of the soft-drink concentrate can be found by using Eq. 4.8 given in the text.

ie $$C_{\text{vit C}} = \frac{\text{Peak height (vit C)} \times C_{\text{caf}}}{\text{Peak height (caf)} \times R_{\text{vit C/caf}}}$$

5.0 mg of caffeine was added to the soft-drink concentrate before diluting it to a volume of 100.0 cm^3. Therefore the concentration of caffeine in the solution analysed was 50.0 mg dm^{-3}

The units mg dm^{-3} have been chosen as the answer is required in these terms.

$$\therefore \quad C_{\text{vit C}} = \frac{6.95 \times 50.0}{8.30 \times 0.74} \text{ mg dm}^{-3}$$

$$= 57 \text{ mg dm}^{-3}$$

Similarly $$C_{\text{sac}} = \frac{\text{Peak height (sac)} \times C_{\text{caf}}}{\text{Peak height (caf)} \times R_{\text{sac/caf}}}$$

$$= \frac{5.00 \times 50.0}{8.30 \times 1.29} \text{ mg dm}^{-3}$$

$$= 23.3 \text{ mg dm}^{-3}$$

The results obtained relate to the concentration of vitamin C and of saccharin in the diluted sample analysed. This was prepared by diluting 20.0 cm^3 of the soft drink concentrate to 100.0 cm^3 ie a fivefold dilution. Therefore the concentrations in the actual sample were;

$$\text{Vitamin C} = 5 \times 57 \text{ mg dm}^{-3}$$

$$= 285 \text{ mg dm}^{-3}$$

$$\text{Saccharin} = 5 \times 23.3 \text{ mg dm}^{-3}$$

$$= 116 - 117 \text{ mg dm}^{-3}$$

If you got both of these correct, well done.

****

SAQ 8.2a

When using

(*A*) the single internal-standard method,

(*B*) the multiple internal-standard method,

state which of the following are assumed to be constant:

(*i*) the ratio of the measured physical parameters,

(*ii*) the ratio of the detector responses given by the analyte and the internal standard,

(*iii*) the concentration of the analyte,

(*iv*) the concentration of the internal standard.

Response

(*A*) The single internal-standard method

Let us first remind ourselves of the fundamental equation for use with the internal standard method. This is

$$\frac{Y_1}{Y_2} = R\frac{C_1}{C_2}$$

where Y_1 and Y_2 are the measured parameters,
C_1 and C_2 are the concentrations of the analyte and internal standard respectively.
R is the detector response ratio.

(*i*) The ratio Y_1/Y_2 **is not assumed** to be constant as both Y_1 and Y_2 are the measured parameters obtained from the instrumental method of analysis used and therefore are related to the concentration of the analyte (C_1) and internal standard (C_2) respectively.

(*ii*) When using the single internal standard method, the value of R must be assumed to be a constant, and must be of a known value. By assuming R to be constant we are assuming that a linear relationship exists between Y_1 and C_1 and between Y_2 and C_2. R is therefore in simple terms, a ratio between the slopes of these 2 lines.

(*iii*) In no way can the concentration of the analyte be assumed to be constant, as the whole idea of the analysis is to determine this figure.

(*iv*) When carrying out a number of analyses by using the single internal standard method, the analyst may well maintain the concentration of the internal standard constant. However there is no need to do this, and nor do we need to assume that this value is constant.

Therefore when using the single internal-standard method, only the value of R (detector response ratio) must be assumed to be constant.

(*B*) The multiple internal-standard method

(*i*) As in part (*A*) of this answer Y_1/Y_2 is not assumed to be constant.

(*ii*) You must remember, that when using the multiple internal standard method, a graph will be plotted of the ratio of the measured parameters (Y_1/Y_2) against analyte concentration (C_1). Usually when this method is used, the resultant graph will be a straight line. However there are instances where the graph may be curved over part or all of the concentration range examined. This in no way invalidates the procedure, provided sufficient data points have been used accurately to predict the shape of the curve. As the slope of the curve at any one time is a function of R/C_2 (see Eq. 8.7), and assuming that the value of C_2 (concentration of the internal standard) remains constant, then any alteration in the slope of the curve must be due to a change in the value of R.

Therefore we cannot necessarily assume R to be constant.

(*iii*) See answer to part (*A*) (*iii*) of this question

(*iv*) It is normal practice when using the multiple internal standard method to use a constant concentration of the internal standard (C_2). We can then plot a graph of the ratio of the measured parameters (Y_1/Y_2) against the concentration of the analyte (C_1). The slope of the graph at any particular point is given by the value R/C_2 which is often written as R'. Therefore it is correct to say that the concentration of the internal standard is held constant.

****

SAQ 8.3a

Select from the following list of compounds, those which you consider *might be useful* as internal standards for the glc analysis of an aqueous effluent containing traces of butan-2-one ($C_2H_5COCH_3$).

Hexane	**(C_6H_{14})**
Pentan-2-one	**($CH_3CH_2CH_2COCH_3$)**
Propanone	**(CH_3COCH_3)**
Toluene	**($C_6H_5CH_3$)**
Acetyl chloride (ethanoyl chloride)	**(CH_3COCl)**
Propan-2-ol	**($CH_3CH(OH)CH_3$)**

Response

Although you may not be fully familiar with glc as an analytical technique, by referring to examples quoted in this Unit and to the ensuing discussion in Section 8.1 you should have been able to suggest from the list of compounds quoted, which of them might be useful as an internal standard for the analysis of an aqueous effluent containing traces of butan-2-one. The properties you should have considered when making your choice were as follows.

— *The internal standard should not already be present in the analyte sample.*

As no information was given as to the other compounds present in the aqueous effluent, we can only assume that those quoted in the list were absent.

— *The internal standard must be separable from the analyte, and all other compounds present in the sample.*

Because they are homologues of the analyte, the only two compounds that you can definitely say will be separable under optimum experimental conditions are propanone and pentan-2-one. The other compounds may well be easily separable, but we have no way of knowing this to be true.

— *The internal standard must not react chemically with the sample.*

One of the compounds listed would react with the analyte solution. Acetyl chloride reacts violently with water according to the following equation,

$$CH_3COCl + H_2O \rightarrow CH_3COOH + HCl$$

and therefore could not be considered as a possible internal standard.

— *The internal standard must be miscible with the solution.*

Hexane and toluene are both hydrocarbons and are therefore likely to be only sparingly soluble in water. Thus simply on grounds of miscibility they should not be considered as internal standards.

Of the compounds listed therefore propanone and pentan-2-one appear to offer themselves as good internal standards. Remember however that propanone (acetone) is very volatile and this could lead to a reduction in the overall accuracy of the procedure.

Propan-2-ol might be useful as an internal standard. It is miscible with water in all proportions, but we should need to investigate whether or not it is separable from the analyte, and all other constituents of the sample, before it could be chosen.

The answer you have given should therefore have included propanone, pentan-2-one and possibly propan-2-ol, and should definitely have excluded hexane, toluene, and acetyl chloride.

****

SAQ 9.4a

The figure below, represents a block diagram of an hplc system interfaced to a microcomputer, which forms a data station within your laboratory.

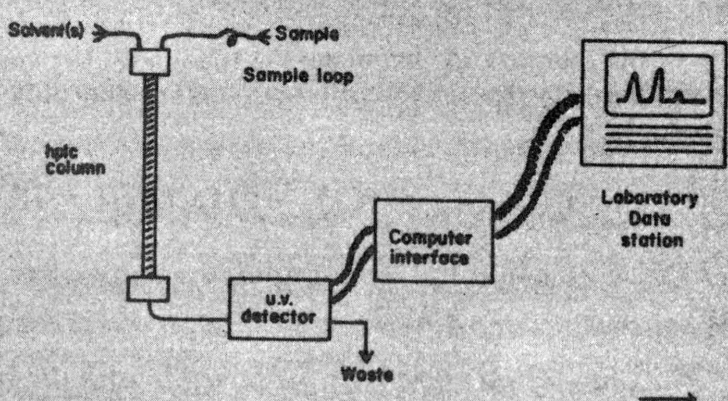

→

SAQ 9.4a (cont.)

When analysing a mixture containing a number of components by hplc, the individual components eluted from the hplc column are passed into the U.V detector. The resulting absorbance signal from the detector is passed *via* a suitable interface into the microcomputer forming the laboratory data station.

In the mixture to be analysed there is only a single analyte to be determined, and the technique of single standard-addition is to be used. You may assume that no dilution of the analyte occurs when preparing the sample containing the standard added.

(*i*) What information will the data station require to have stored, before the analysis can be carried out and the result presented in concentration terms?

(*ii*) What information will the operator need to provide to the microcomputer?

(*iii*) What information will the hplc instrument need to provide to the microcomputer?

(*iv*) What functions does the microcomputer have to perform in this analysis?

Response

Before we consider the computing aspect of this analysis we need to revise the requirements for the use of the single standard-addition method.

Eq. (7.5) in Section 7 of this Unit states:

$$C = \frac{Y_O C_S}{Y_1 - Y_O}$$

which can be rewritten for this example as

$$C = \frac{(\text{Area})_O C_S}{(\text{Area})_1 - (\text{Area})_O}$$

where C = concentration of the analyte to be determined,

C_S = increase in concentration of the analyte caused by the addition of the standard,

$(\text{Area})_O$ = initial area underneath the analyte peak,

$(\text{Area})_1$ = area under the analyte peak after the addition of the standard.

(*i*) There are three essential pieces of information the computer will need to have stored before we commence with the analysis.

- It must be capable of identifying the analyte amongst all the other components present in the mixture. This will be done on a time basis, measured from the time when the sample was introduced onto the hplc column.

- It must be capable of integrating the area under a chromatographic peak.

- It must be capable of performing the calculation.

If this analysis is performed routinely, then the value of C_S may be constant. If this is so, then this will also be information stored by the computer.

(*ii*) There are three essential facts that the operator will have to provide.

- Sample identification – details of each sample must be given to the computer before the analysis is commenced.

- The value of C_S, if this is not already known [see answer to part (*i*)].

- The time when each sample was introduced onto the hplc column. This will not normally be in hours and minutes, but will be some form of electrical signal to indicate the start of the analysis. With a laboratory data station one can assume that the computer is already notified of the date.

(*iii*) The hplc instrumentation needs to provide the computer with only one continuous piece of information *viz* the measurement of absorbance as the eluate from the hplc column passes through the detector.

(*iv*) In this example the computer is performing two functions: communication and data processing. These are listed in more detail below.

- Being a laboratory data station, the computer will have to inform the operator that a channel for data acquisition is available.

- The computer will have to identify the analyte present in the mixture, based upon the stored time parameters.

- Having identified the analyte the computer will have to integrate the area under the chromatographic peak in both chromatograms

- The computer will have to perform the calculation in accordance with the formula stored in its memory.

- The information output by the computer will include

- analysis title
- date when the analysis was performed,
- analytical conditions,
- concentration of analyte.

This can in no way be a complete answer, as all computer systems differ from one another to some extent. However, if you managed to get most of these correct, then you have obviously managed to appreciate the interactive role played by the analytical instrument, the computer, and the analyst within the modern analytical laboratory.

****

SAQ 10.a

The following list of statements [(i) to (v)], relate to a comparison of two methods of analysis,

(*A*) the single internal-standard method,

(*B*) the method involving the bracketing of standards around the sample.

(*i*) Both these methods require the existence of a linear relationship between analyte concentration and detector response over the range of concentrations examined.

Answer: true, false, or 'insufficient information given'.

(*ii*) Both methods increase the accuracy of a determination.

Answer: true, false, or maybe ⟶

SAQ 10.a (cont.)

(*iii*)	**Method (*A*) is more rapid to carry out than method (*B*).**
Answer:	**true, false, or 'insufficient information given'.**
(*iv*)	**Both methods are capable of overcoming interference effects caused in the analysis by another component in the analyte solution or sample.**
Answer:	**true or false.**
(*v*)	**Both methods are suitable for regular routine analysis.**
Answer:	**true, false, or maybe.**

Response

(*i*) The correct answer is *false*

Although the internal standard method certainly relies on the existence of a linear relationship between analyte concentration and detector response, the method involving the bracketing of standards around the sample does not. Provided the standards have concentrations very close together, and on either side of the analyte concentration, the actual curve in the calibration plot may then approximate to a straight line.

(*ii*) The statement is 'both methods increase the accuracy of a determination' – but increase over what? As we have not been given this information, the correct answer must be *maybe*.

It is true that in normal circumstances the use of the bracketing principle will increase the accuracy of a determination over that obtained for a calibration plot. The internal-standard method on the other hand may or may not give increased accuracy but should give an increase in precision over results obtained by using a calibration plot.

(*iii*) In order to use the single internal-standard method we must initially calculate the value of the response factor. Each subsequent sample analysed then requires the addition of an internal standard. With the bracketing of standards method, we must first produce a calibration plot and use that plot to estimate the concentration of analyte in our sample. Only when we have done this can we prepare fresh standards to bracket the sample. The correct answer is therefore *true* – particularly when the initial response factor has been found.

(*iv*) The correct answer is *false*

The methods on their own cannot in general compensate for interfering effects occurring within the sample solution. There are a few exceptions however, arc emission spectroscopy for instance, where the use of an internal standard will compensate for alterations in experimental conditions. This may or may not be classed as an interference effect.

(*v*) Because both methods require initial sample pre-treatment or the preparation of additional standards, an alternative procedure for routine analysis should be adopted if at all possible. Where however, an alternative method has not been shown to be suitable, then a procedure involving one of these methods could be used. The answer is therefore *maybe*.

If you did not get all these parts correct I suggest that you revise the appropriate sections in this Unit before re-attempting the questions.

****

SAQ 10.1a

> **Do you consider the method of preparing gaseous mixtures by weight would be appropriate to prepare 2 litres of a 10% v/v mixture of CO_2 and nitrogen?**

Response

We need first to calculate the weight of CO_2 that will be added to the container.

Now 10% of 2.00 dm^3 = 200 cm^3, and

22.4 dm^3 of CO_2 at STP weighs 44 g

$$\therefore \quad 200\ cm^3 \text{ of } CO_2 \text{ at STP weighs } \frac{200}{22.4 \times 1000}\ 44\ g$$

$$= 0.39\ g$$

[Relative molar mass of CO_2 = 44]

Now although we can weigh this amount accurately on an analytical balance, analytical balances are not capable of also weighing a container, 2 dm^3 in volume.

However some of the modern 'top pan' balances are capable of weighing to 2 kg with an accuracy of ± 0.01 g. If one of these is available then the gravimetric method becomes feasible.

****

SAQ 10.1b

> **A plastic bag holds 1750 cm^3 of gas at atmospheric pressure. If 100 μl of vinyl bromide (M_r = 116) is introduced into the bag and the pressure of gas in the bag is adjusted to 1.00 atmosphere, calculate the percentage by volume of vinyl bromide in the gaseous atmosphere.**
>
> **You may assume all the vinyl bromide inside the bag to be vapourised.**
>
> **Temperature of the gas = 20 °C,**
> **Barometric pressure = 1.00 atmosphere,**
> **Density of vinyl bromide = 0.95 g cm^{-3}.**

Response

We need to determine initially the volume that 1 mole of vinyl bromide would occupy at 20 °C and 1 atmosphere pressure.

As a gas expands on being heated the volume at STP of 22.4 dm^3 will have increased in the ratio 293/273.

$\therefore$ Vol occupied by 1 mole at 20 °C (293 K)

$$= 22.4 \times \frac{293}{273} \text{ dm}^3$$

$$= 24.04 \text{ dm}^3$$

Vinyl bromide has a density of 0.95 and therefore 116 g of liquid vinyl bromide (1 mole) will occupy $\frac{116}{0.95}$ cm^3

$$= 122.1 \text{ cm}^3$$

We now need to convert 100 μl to cm^3

$$100\ \mu l = 0.100\ cm^3$$

Therefore, the volume occupied by 0.100 cm^3 of liquid vinyl bromide in the gaseous state can be calculated as follows:

122.1 cm^3 of vinyl bromide occupies 24.04 dm^3

$\therefore$ 0.100 cm^3 of vinyl bromide occupies

$$\frac{0.100}{122.1} \times 24.04\ dm^3 = 0.0197\ dm^3$$

$$= 19.7\ cm^3$$

The total volume of gas inside the bag at the pressure of 1 atmosphere is 1750 cm^3.

$\therefore$ % v/v of vinyl bromide in the gaseous mixture is

$$\frac{19.7}{1750} \times 100\%$$

$$= 1.1\%\ v/v$$

If you managed to get the correct answer, well done.

****

SAQ 10.1c

Designate as 'true' or 'false' the following statements relating to the preparation of gaseous mixtures. If you cannot decide upon an answer, return to the text, as it will be of no value to you to hazard a guess. ⟶

SAQ 10.1c (cont.)

(*i*) **The gravimetric method of producing gas mixtures is well suited to the preparation of cylinders of mixed gases at high pressures.**
True or false?

(*ii*) **The preparation of gaseous mixtures by measuring the pressure increase inside an evacuated cylinder, is a good example of a dynamic method.**
True or false?

(*iii*) **Most plastic materials are suitable for the preparation and storage of mixtures of gases**
True or false?

(*iv*) **Permeation tubes are good devices for the preparation at ambient temperatures, of low-concentration mixtures of relatively low volatility organic compounds.**
True or false?

Response

(*a*) The correct answer is *true*

The gravimetric method is an accurate way of preparing large quantities of gas mixture, generally at concentrations greater than 1% of the analyte gas. Concentrations lower than 1% can also be prepared, but generally by dilution of a more concentrated mixture.

(*b*) The correct answer is *false*

This is not an example of a dynamic method, but of a static method for preparing gas mixtures. Dynamic methods are those involving some form of continuous mixing of the analyte gas and the diluent.

(*c*) The correct answer is *false*

Most plastics are permeable to gases, particularly if the gases are of small molecular size. There are very few plastics that are capable of restricting permeation – nylon and teflon have proved useful in the preparation of some gas mixtures.

(*d*) The correct answer is *false*

While permeation tubes are excellent for the preparation of dilute gas mixtures, the compound being diluted must be of high volatility, so that it is under pressure within the tube itself. A compound of relatively low volatility would not diffuse to any useful extent through the plastic tube into the gas stream passing over it.

****

SAQ 11.1a

What conclusions could you draw, if when analysing a sample of a certified reference material by your own laboratory procedure, you obtained a result different from that of the certified value?

Response

There are a number of possible conclusions that can be drawn, the main ones being as follows:

– That the method being used is not appropriate for the analysis of the reference material. Remember that the reference material being used is possibly not a pure material, and thus the other substances present may well interfere with your method.

— Your calibration standards are at fault. If this is so, it is best to recalibrate with fresh standards, or possibly try alternative quantitative methods ie standard addition or internal standard methods.

— You are a poor practitioner!

****

SAQ 11.2a

How do you consider we arrived at a value of 98 → 102% as an acceptable range for recovery?

Response

Earlier in this Unit we considered the accuracy which is normally attainable when using a relative method of analysis. The figure quoted was $\pm 1 \rightarrow 2\%$ relative. Relating this to our percentage recovery figure therefore gives us an acceptable spread of results of

$$100 \pm 2\%.$$

ie $$98 \rightarrow 102\%$$

****

SAQ 11.4a

Two analysts are asked to analyse the same sample by different procedures, each performing the analysis 10 times. If they obtain mean values which are significantly different from one another, but at the same time obtain results with equivalent precision, which of the following conclusions can we draw from their results? ⟶

SAQ 11.4a (cont.)

> **(*i*) One of the methods contains an indeterminate error.**
>
> **(*ii*) Each method contains an indeterminate error.**
>
> **(*iii*) One of the methods contains a determinate error.**
>
> **(*iv*) Each method contains a determinate error.**
>
> **(*v*) One of the results is inaccurate.**
>
> **(*vi*) Both results are inaccurate.**

Response

In order to answer parts (*i*) to (*iv*) we need to refer back to Part 2 of this unit where definitions of determinate and indeterminate error were given.

In any set of replicate results, a spread of values will be obtained, this being due to the *indeterminate error* which is always present in any type of analysis. Therefore conclusion (*i*) is untrue, and conclusion (*ii*) is true, both of the methods *do* possess indeterminate errors.

Determinate errors on the other hand are errors that can be defined and in theory have a measurable value. Now bearing in mind that both methods of analysis are of equal precision, but produce differing mean values, *at least one of the methods* must contain a determinate error. Both of the methods may contain a determinate error, but this is not a conclusion that we can draw at this stage. Thus conclusion (*iii*) is correct, but we have insufficient evidence to support conclusion (*iv*) as also being correct. As we have not been told the true analyte concentration, then a similar argument must apply to part (*v*) and (*vi*) as is used in part (*iii*) and (*iv*).

One of the results must be wrong, but both of the answers may be wrong. Conclusion (*v*) is therefore true but again we have insufficient evidence to support conclusion (*vi*) as also being true.

Units of Measurement

For historic reasons a number of different units of measurement have evolved to express quantity of the same thing. In the 1960s, many international scientific bodies recommended the standardisation of names and symbols and the adoption universally of a coherent set of units—the SI units (Système Internationale d'Unités)—based on the definition of five basic units: metre (m); kilogram (kg); second (s); ampere (A); mole (mol); and candela (cd).

The earlier literature references and some of the older text books, naturally use the older units. Even now many practicing scientists have not adopted the SI unit as their working unit. It is therefore necessary to know of the older units and be able to interconvert with SI units.

In this series of texts SI units are used as standard practice. However in areas of activity where their use has not become general practice, eg biologically based laboratories, the earlier defined units are used. This is explained in the study guide to each unit.

Table 1 shows some symbols and abbreviations commonly used in analytical chemistry; Table 2 shows some of the alternative methods for expressing the values of physical quantities and the relationship to the value in SI units.

More details and definition of other units may be found in the *Manual of Symbols and Terminology for Physicochemical Quantities and Units*, Whiffen, 1979, Pergamon Press.

Table 1 *Symbols and Abbreviations Commonly used in Analytical Chemistry*

Symbol	Meaning
Å	Angstrom
A_r(X)	relative atomic mass of X
A	ampere
E or *U*	energy
G	Gibbs free energy (function)
H	enthalpy
J	joule
K	kelvin (273.15 + *t* °C)
K	equilibrium constant (with subscripts p, c, therm etc.)
K_a,K_b	acid and base ionisation constants
M_r(X)	relative molecular mass of X
N	newton (SI unit of force)
P	total pressure
s	standard deviation
T	temperature/K
V	volume
V	volt ($J\ A^{-1}\ s^{-1}$)
a, *a*(A)	activity, activity of A
c	concentration/ mol dm^{-3}
e	electron
g	gramme
i	current
s	second
t	temperature / °C
bp	boiling point
fp	freezing point
mp	melting point
≈	approximately equal to
<	less than
>	greater than
e, exp(*x*)	exponential of *x*
ln *x*	natural logarithm of *x*; ln *x* = 2.303 log *x*
log *x*	common logarithm of *x* to base 10

Table 2 *Alternative Methods of Expressing Various Physical Quantities*

1. **Mass (SI unit : kg)**

$$g = 10^{-3}\ kg$$
$$mg = 10^{-3}\ g = 10^{-6}\ kg$$
$$\mu g = 10^{-6}\ g = 10^{-9}\ kg$$

2. **Length (SI unit : m)**

$$cm = 10^{-2}\ m$$
$$Å = 10^{-10}\ m$$
$$nm = 10^{-9}\ m = 10Å$$
$$pm = 10^{-12}\ m = 10^{-2}\ Å$$

3. **Volume (SI unit : m^3)**

$$l = dm^3 = 10^{-3}\ m^3$$
$$ml = cm^3 = 10^{-6}\ m^3$$
$$\mu l = 10^{-3}\ cm^3$$

4. **Concentration (SI units : mol m^{-3})**

$$M = mol\ l^{-1} = mol\ dm^{-3} = 10^3\ mol\ m^{-3}$$
$$mg\ l^{-1} = \mu g\ cm^{-3} = ppm = 10^{-3}\ g\ dm^{-3}$$
$$\mu g\ g^{-1} = ppm = 10^{-6}\ g\ g^{-1}$$
$$ng\ cm^{-3} = 10^{-6}\ g\ dm^{-3}$$
$$ng\ dm^{-3} = pg\ cm^{-3}$$
$$pg\ g^{-1} = ppb = 10^{-12}\ g\ g^{-1}$$
$$mg\% = 10^{-2}\ g\ dm^{-3}$$
$$\mu g\% = 10^{-5}\ g\ dm^{-3}$$

5. **Pressure (SI unit : N m^{-2} = kg m^{-1} s^{-2})**

$$Pa = Nm^{-2}$$
$$atmos = 101\ 325\ N\ m^{-2}$$
$$bar = 10^5\ N\ m^{-2}$$
$$torr = mmHg = 133.322\ N\ m^{-2}$$

6. **Energy (SI unit : J = kg m^2 s^{-2})**

$$cal = 4.184\ J$$
$$erg = 10^{-7}\ J$$
$$eV = 1.602 \times 10^{-19}\ J$$

Table 3 *Prefixes for SI Units*

Fraction	Prefix	Symbol
10^{-1}	deci	d
10^{-2}	centi	c
10^{-3}	milli	m
10^{-6}	micro	μ
10^{-9}	nano	n
10^{-12}	pico	p
10^{-15}	femto	f
10^{-18}	atto	a

Multiple	Prefix	Symbol
10	deka	da
10^{2}	hecto	h
10^{3}	kilo	k
10^{6}	mega	M
10^{9}	giga	G
10^{12}	tera	T
10^{15}	peta	P
10^{18}	exa	E

Table 4 *Recommended Values of Physical Constants*

Physical constant	Symbol	Value
acceleration due to gravity	g	$9.81\ \mathrm{m\ s^{-2}}$
Avogadro constant	N_A	$6.022\ 05 \times 10^{23}\ \mathrm{mol^{-1}}$
Boltzmann constant	k	$1.380\ 66 \times 10^{-23}\ \mathrm{J\ K^{-1}}$
charge to mass ratio	e/m	$1.758\ 796 \times 10^{11}\ \mathrm{C\ kg^{-1}}$
electronic charge	e	$1.602\ 19 \times 10^{-19}\ \mathrm{C}$
Faraday constant	F	$9.648\ 46 \times 10^{4}\ \mathrm{C\ mol^{-1}}$
gas constant	R	$8.314\ \mathrm{J\ K^{-1}\ mol^{-1}}$
'ice-point' temperature	T_{ice}	273.150 K exactly
molar volume of ideal gas (stp)	V_m	$2.241\ 38 \times 10^{-2}\ \mathrm{m^3\ mol^{-1}}$
permittivity of a vacuum	ϵ_o	$8.854\ 188 \times 10^{-12}\ \mathrm{kg^{-1}\ m^{-3}\ s^4\ A^2}$ ($\mathrm{F\ m^{-1}}$)
Planck constant	h	$6.626\ 2 \times 10^{-34}\ \mathrm{J\ s}$
standard atmosphere pressure	p	$101\ 325\ \mathrm{N\ m^{-2}}$ exactly
atomic mass unit	m_u	$1.660\ 566 \times 10^{-27}\ \mathrm{kg}$
speed of light in a vacuum	c	$2.997\ 925 \times 10^{8}\ \mathrm{m\ s^{-1}}$